THE STEEL INDUSTRY AND THE ENERGY CRISIS

THE STEEL INDUSTRY AND THE ENERGY CRISIS

edited by JULIAN SZEKELY

Center for Process Metallurgy
The State University of New York at Buffalo
Buffalo, New York

Proceedings of the Fourth C. C. Furnas Memorial Conference

MARCEL DEKKER, INC. New York

MARCEL DEKKER, INC.

270 Madison Avenue, New York 10016

LIBRARY OF CONGRESS CATALOG CARD NUMBER: 74-30737

ISBN: 0-8247-6270-3

Current printing (last digit):
10 9 8 7 6 5 4 3 2 1

PRINTED IN THE UNITED STATES OF AMERICA

PREFACE

This volume contains the proceedings of the Fourth C. C. Furnas Memorial Conference, held in Buffalo, New York on November 11-12, 1973. These conferences are held annually to commemorate the late Clifford C. Furnas, a noted scientist, statesman and educator, the first president of the State University of New York at Buffalo. In selecting the theme and the detailed program for the meeting a deliberate effort is made to promote a free interchange of ideas between steel mill operators, researchers and academics.

The theme of this Conference, "The Steel Industry and the Energy Crisis," was thought to be very topical because of the political and economic developments in 1972 and 1973. These events, which led to a very drastic alteration in both the availability and the price structure of oil and natural gas, have forced a major reassessment in the energy consumption patterns of all the industrialized nations in the western world.

In the United States the steel industry accounts for about 5 percent of the total energy consumption, while the corresponding figure in Japan is a staggering 20 percent. Western European nations occupy an intermediate position between these two extremes. The drastic increase in the price of oil and of natural gas, coupled with the restricted availability of natural gas in many locations, has hit the steel industry particularly hard. This is due, in part, to recent trends which have indicated a particularly rapid increase in the consumption of these fuels and at the expense of coking coal.

The energy crisis with which the steel industry is faced demands both short-term and long-term responses. In the short term, measures have to be taken to effect fuel economies. Moreover, energy requirements and fuel-cost effectiveness are becoming important and overriding factors in the selection of alternatives from <u>available</u> processes. In the long term, the energy crisis requires the development of new, more efficient metal-producing systems, possibly through the use of nuclear energy.

The contributions contained in this volume reflect both short-term and long-term industrial responses to the energy crisis. These papers range from an authoritative statement of the problem to the discussion of very practical steps that may be taken in a given plant to effect fuel economy. Of particular interest may be two contributions, the one dealing with the experiences and future plans of the Japanese steel industry, the other discussing the prospects of nuclear steelmaking.

The editor's task is always very difficult in the compilation of a volume written by contributors with highly diverse backgrounds and interests. No attempt was made to enforce a uniform style or even a uniform format of presentation. This may have tempered the diversity of the contributions and thus some of the real sense of the meeting would have been lost.

In closing it is a pleasure to thank all the contributors for their work, Mrs. C. C. Furnas for her encouragement and to express a formal appreciation to Dr. McAllister Hull, dean of the Graduate School, State University of New York at Buffalo, who officiated at the opening of the Conference.

The organization of the conference and the production of this volume was made possible by a grant from the A. E. Anderson Foundation.

Finally, the editor wishes to express his thanks to Mrs. Lucille Delmar for her help in organizing all phases of this work and to Mrs. Christine Jerome for typing the manuscript.

Buffalo, New York Julian Szekely

LIST OF CONTRIBUTORS

R. W. Bouman, Section Manager, Homer Research Laboratories, Bethlehem Steel Corporation

A. C. Elliott, Superintendent, Utilities Department, The Steel Company of Canada, Ltd.

John F. Elliott, Professor of Metallurgy, Massachusetts Institute of Technology

Frank W. Luerssen, Vice President -- Research, Inland Steel Company

Mikio Takagi, General Manager, Energy Control Department, Nippon Steel Corporation

J. L. Uvira, Research Technical Advisor, The Steel Company of Canada, Ltd.

Werner Wenzel, Professor, Institut für Eisenhüttenwesen Technische Hochschule

CONTENTS

THE STEEL INDUSTRY AND THE ENERGY CRISIS

PERSPECTIVES: ENERGY AND THE U.S. STEEL INDUSTRY

Frank W. Luerssen

Vice President - Research
Inland Steel Company
East Chicago, Indiana

It is an honor and pleasure to make an initial statement to this Fourth C.C. Furnas Memorial Conference which today is addressing itself to an analysis of the possible near and longer term responses of the steel industry to the important energy problems it now faces.

It is my assignment and intention to present some broad perspectives on the energy problem-steel industry interface. The distinguished speakers who follow on this program and the discussions scheduled surely will illuminate in greater measure the real energy issues facing the industry in the days and years ahead. The objective of this paper will have been achieved if some of the important aspects of the problem are highlighted in ways that will not substantially add to that which is already a voluminous confusion of fact, near fact, and opinion.

It was clearly evident to the careful observer that only a year ago many local suppliers of energy seemed not to have appreciated the importance of the developing energy supply dislocations as respects their own businesses and the public, and some industrial firms inexcusably were completely oblivious to the problem as recently as a few months ago. It seems that brown-outs and gasoline and heating oil shortages indeed have done more than any publication, presidential message, or the like to communicate the seriousness of the problem to the public. How much more difficult it will be to convey to the public a credible story of the potentially serious impact on the steel industry of the arbitrary energy allocations that will surely emerge in the absence of a clear public energy policy. One of the great challenges facing this industry right now is to generate a scholarly and wholly defensible position on the energy issue and to annunciate that position effectively in places where it counts. There is a growing awareness in the industry that this must be done. The organizers of this conference are to be commended for providing this forum for discussing these issues in a relatively unbiased climate.

Examine now some clear signs which pretold of the existence of an emerging energy problem.

Inactivity, New Crude Oil Production (depletion allowances)
Inactivity, New Gas Transmission Lines (price controls)
Decline, Coal Mining Activity (health, safety, environment)

The underlying features of these past situations in the energy-supplying industries, shown in parentheses, are representative of some of the same internal and external pressures currently being experienced by the steel industry and work to the same end; namely, curtailment of production.

It is no news to this or any other informed group that much of the industry is caught in the grasp of a severe capital squeeze in the face of a strong demand for its products. Superimposed on capital needs for major maintenance, modernization, and expansion are needs to meet health, safety, and environmental obligations. For example, estimates of the oil industry capital needs to meet new production and petroleum distribution requirements have been placed at 150 billion dollars over the next 10 years. The estimate for power generation facilities is in excess of 200 billion dollars for the next decade. Capital demands in steel will aggregate without question something close to 35 billion of 1973 dollars, or roughly triple that of the past 10 years based on 1963 dollars. This must be done in the face of what is predicted to be a booming world steel demand, an intensification of the already short supply and rising costs of iron-bearing materials, and price controls which are forcing domestic suppliers to seek higher margins from selling in foreign markets. Clearly, many factors are severely limiting steel's ability to generate the profits necessary to make the capital expenditures required. Can the required rate of capital formation be achieved? Perhaps, but only under the best of circumstances.

Without doubt there will be important implications of this capital crisis on the nature of feasible responses available to the steel industry to meet its segment of the energy problem. Every advantage must be taken of opportunities to combine whatever investments must be made to resolve our energy problems in concert with our central capital programs for expansion and modernization. Solutions must be thoroughly explored that do not impose on our industry the massive capital expenditures for energy conversions such as would be the case with captive coal gasification plants. Most importantly, the industry must avoid transient solutions that will lead to the future compounding of capital requirements.

A brief overview of the national energy situation seems appropriate before its consequences for the steel industry are examined. With only 6% of the world's population, the United States consumes nearly one-third of the world's energy supply. United States consumption reached about 72 quadrillion Btu in 1972, having doubled in the last 20 years, and is probably due to double again in the next 15 years. Natural gas production has or will very soon peak out. It is said that our petroleum needs will double in the next 15 years with all of this growth coming from imports. These imports will represent more than 50% of supply. There will be a continuing need to consume high sulfur residual oils by environmentally acceptable schemes such as blast furnace injection. Coal is and will continue to be available, although low sulfur steaming coals and coals suitable for producing metallurgical coke may be in short or marginal supply. Nuclear power generation will grow rapidly during the same period. On this basis we should anticipate strict limitations on the uses of and sharply higher prices for natural gas; periodic and at times serious short-term supply problems with and higher prices for fuel oil; a relatively easy situation with regard to coal supplies, although higher prices seem to be inevitable; and, a greater reliance on nuclear energy.

The steel industry now consumes about 5% of the nation's energy supply, employing on the average about 25 million Btu per raw steel ton of production. Table 1 shows the magnitudes of the energy uses by source for 1972 and the author's estimate for 1973.

TABLE 1.

Steel Industry Energy Uses
(Actual 1972 and Estimated 1973)

Energy sources	Amount	% of Total (Btu basis)	Amount	% of Total (Btu basis)
Coal	81.4 MMNT	63.0	90.4 MMNT	64.5
Fuel oil	1313 MM gal	6.0	2000 MM gal	8.4
Natural gas	772 BCF[a]	24.0	725 BCF	20.6
Purchased power	39.4 BkWh	4.1	43.5 BkWh	4.2
Other	94 TBtu	2.9	94 TBtu	2.6
Total		100.0		100.0
Raw steel production (Millions of net tons)		133.2		151.8

[a]A possible positive error of some magnitide exists in the natural gas consumption as reported by AISI.

Note that the industry even now is juggling available energy sources to meet its needs for higher production levels while compensating for cutbacks in natural gas supply. Most notable is the whopping 50% increase in fuel oil consumption reaching an all-time peak level for the industry. While there is some speculation on the accuracy of the 1973 predictions, the data clearly indicates that there is a fair flexibility as regards the substitution of fuel oil for natural gas, which probably comes as no surprise to this informed audience.

To explore the potential convertibility of fuel uses in greater detail, a computation was made of the shifts that could have been made in energy usage from gas to oil and oil to coal during 1972. This was done on the basis of the application of known technology, no major new capital investment, and maintaining an acceptable environmental program by current standards. While this analysis represents an extrapolation of the situation in my own company with all the weakness inherent in that approach, the results shown in Table 2 are worth discussing at least within the broad framework of my assignment.

TABLE 2.

Steel Industry Estimated Convertibility of Fuel Usage - 1972 (Basis: Inland Steel Company Energy Model)

Energy sources	1972 AISI data	1972 Converted	Change (TBtu)
Coal	81.4 MMNT	84.4 MMNT	+ 75
Fuel oil	1313 MM gal	2982 MM gal	+248
Natural gas	772 BCF	456 BCF	+75

Estimated least capital cost - $200 Million

The results seem to confirm current observations, that there is good convertibility of gas to oil usage in the near term at modest capital cost but relatively poor convertibility of these fuels to coal. If in the near term the industry is severely restricted in its consumption of both gas and oil, the consequence must be a reduction in steel production. It now appears certain that there will be both fuel oil and natural gas restrictions in the intermediate term as well as in the long term. Therefore the response of the industry must be pointed toward increased use of coal, of energy derived from coal, and of nuclear energy. Furthermore, plans should be implemented to move in those directions as soon as practicable.

The massive programs required to move in this direction will involve the development of new raw materials operations, new provisions for raw materials transport, the construction of coke plants and blast furnace plants, accelerated programs for phasing out open hearth steelmaking in favor of BOF and electric furnaces, and the electrification of heating operations. This would seem to impose staggering economic burdens on the industry for which it is not now prepared and would most certainly seriously reduce the industry's capability to expand to meet projected market demands.

The situation is not all that dark because there exist approaches for improving our energy utilization efficiency and therby easing, somewhat, our total capital requirements. Such actions as the adoption of the best existing blast furnace technology, modernization of steelmaking facilities and greater reliance on continuous casting will be effective. It is disconcerting, however, to note that energy consumption per ton has not improved over the past three-year period, and there is little likelihood that 1973 results will be substantially better. The industry probably will have consumed 3,720 trillion Btu in 1973 represented by the upper line in Fig. 1 rather than the 3,670 trillion Btu represented by the lower line that might have been expected from a concerted effort at energy conservation. Perhaps this is not an unexpected result since steel producers only now are initiating cohesive energy conservation programs in recognition that a serious problem exists.

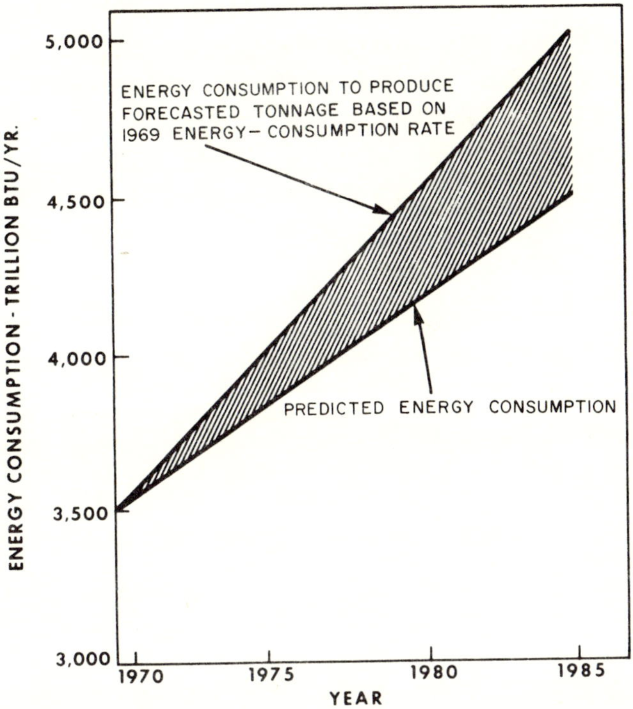

Fig. 1. Forecast energy consumption for the United States States steel industry 1969-1985. From the paper, "Energy and the U.S. Steel Industry", by M. Tenenbaum and F. W. Luerssen, presented at the 5th Annual Conference of the International Iron & Steel Institute in Toronto, Canada, October 11-15, 1971.

While energy conservation programs are important, effective longer range solutions must be sought to the imposing energy restraints that we are now facing. The steel industry and in fact a broad segment of all industry needs the guidance of an energy perspective of national scope on which to base future plans. The only reasonable basis for such policy must be a series of studies of the impact of energy decisions on the industries affected and on the public as a whole. There is much at stake in the steel industry. Surely there must be accommodations among expansion programs, environmental programs and energy programs to enable the steel industry to continue to perform its basic role in our economy and to meet the steel needs of the nation. We must be at the task of promptly illuminating the important issues which must form the basis for planning such accommodations. This conference can and hopefully will work effectively in that direction.

DEFINITION OF TERMS

TBtu 10^{12} Btu

BCF 10^{9} normal cubic feet

M 10^{3}

USES OF ENERGY IN THE PRODUCTION OF STEEL

John F. Elliott

Professor of Metallurgy
Massachusetts Institute of Technology
Cambridge, Massachusetts

From historical times, the cost and availability of fuels in various forms has played an essential role in the technical and economic advancement of the iron and steel industry. The use of coal and coke to replace charcoal saved the British ironmaking industry in the eighteenth century when it no longer was acceptable to consume the forests for the production of charcoal. A marked improvement in blast furnace operations resulted from the invention of the hot blast stove early in the nineteenth century. Also, the use of regenerators made it possible to obtain a high enough temperature for the production of steel in the hearth-type furnace, and the "open hearth" was born in 1864.

The current crisis in the general availability of energy can be expected to have profound effects on the steel industry of the United States. While there will be many difficulties, there are reasons to feel that the industry has adequate capability of meeting these problems. These reasons will be discussed later. As shown in Table 1 (1) the domestic steel industry consumed approximately 5% of all fuel energy used in the United States in 1972. However, the amount of energy actually required for the production of steel domestically is somewhat higher than that shown for several reasons. The first is that the reported figures do not include most of the energy used in the transportation of raw materials, products, etc., to and from the steel plants. Secondly, they do not include the energy used in the production of ores, agglomerates, burnt lime, ferroalloys, etc., which American companies obtain from sources outside their own facilities within the United States. It is difficult to judge the error arising from these omissions, but it probably is not more than a few percent of the reported total energy used by the industry.

TABLE 1

Fuels for Steel Production in the United States in 1973

Fuel	Amount	10^{14} Btu	% of total net
Coal			
for coke	77.1×10^6 tons	20.0	58.0
coke	54.8×10^6 tons	(13.6)	--
coke gas (net)	6.0×10^6 cf	(3.0)	--
oil, tar, etc.	9.3×10^8 gal	(1.4)	--
for general use	4.3×10^6 tons	1.0	2.9
Blast furnace gas (net)	1.5×10^{12} cf	(1.4)	--
Fuel oil	1.3×10^9 gal	1.9	5.5
Natural gas	7.7×10^{11} cf	7.7	22.3
Purchased electricity	3.9×10^{10} kWh	(3.9)	11.3
TOTAL NET		34.5	100.00

Total U.S. energy consumption (1972): 740.0. Average energy per ton of finished steel: 37.6×10^6 Btu.

[a] Total consumption of electricity is 5.17×10^{10} kWh, that is, 485 kWh per ton of finished steel.

There is also some consumption which cannot be identified correctly. The matter will be discussed later.

The data in Table 1 shows that the industry derives over 60% of its energy from coal. The principal secondary fuel obtained from the coal is coke, and the total energy in the coke is approximately 65% of the energy of the total amount of coal consumed. Thus, approximately 40% of the energy consumed by the industry is in the form of coke for processing and smelting of iron ores which are the operations (sintering and the blast furnace) where virtually all the coke is used. Natural gas is the second most important fuel in terms of quantity used in the steel industry. It is noteworthy that the auxiliary fuels arising from the production and use of coke, i.e., coke oven gas, tar, oil, and blast furnace gas, totaled 5.8×10^{14} Btu in 1972. Thus, collectively they constitute the third largest energy source for the industry. The industry consumed a total of 5.2×10^{10} kWh of electricity in 1972. The thermal energy required to produce this electrical energy is estimated to be 5.2×10^{14} Btu.* However, 24% of this electrical energy was produced by the industry and the balance was purchased. The fraction generated at home was obtained by burning some portion of each of the fuels listed in Table 1. The records show that four decades ago, the industry was much more dependent on coal than it is now for its primary source of energy. For example, coal was gasified to produce fuel for open hearth operations and for heating and reheating furnaces. Because of convenience, greater cleanliness and lower cost, oil and natural gas have to a great extent replaced fuels derived from coal in these operations.

USE OF FUELS BY PROCESSES

Discussion of the present and future use of fuels in the production of steel requires that one consider the principal processes

---

*10,000 Btu is considered 1 kWh.

employed by the American steel industry. They are agglomeration processes, coking operations, blast furnace, basic oxygen furnace (BOF), basic open hearth furnace, electric arc furnace, soaking pits, and the hot and cold rolling facilities which include heating, reheating and annealing furnaces, and the rolling mills.

Approximate average patterns of use of fuels in the main process steps for steelmaking on a large scale are shown in Table 2--agglomeration, Table 3--coking, Table 4--blast furnace, Table 5--BOF, Table 6--basic open hearth, Table 7--electric furnace, and Table 8--mills for hot and cold rolling operations. There is rather wide variation in the American industry in the amounts of energy used in each of these operations, principally because of the nature and availability of raw materials. In this discussion, issues associated with whether natural gas, oil, coke oven gas, etc., are used are avoided by considering simply the thermal requirements of a step (see Fig. 1). This is satisfactory because technically these fuels can be interchanged in many places without serious difficulty provided they have low levels of sulfur. The patterns of use of energy given could be simplified even more by showing only the net consumption. However, some additional detail is included; for example, the use of coke oven gas for underfiring coke ovens and blast furnace gas in the blast furnace stoves. A few remarks on some of the more important assumptions follow.

The energy requirements for agglomeration (Table 2) are reasonably typical of the production of moderately basic sinter, and sintering is the principal method for agglomeration in use within an American steelworks . The requirement for pelletizing may be in the range shown and can be somewhat less depending on the particular method employed.

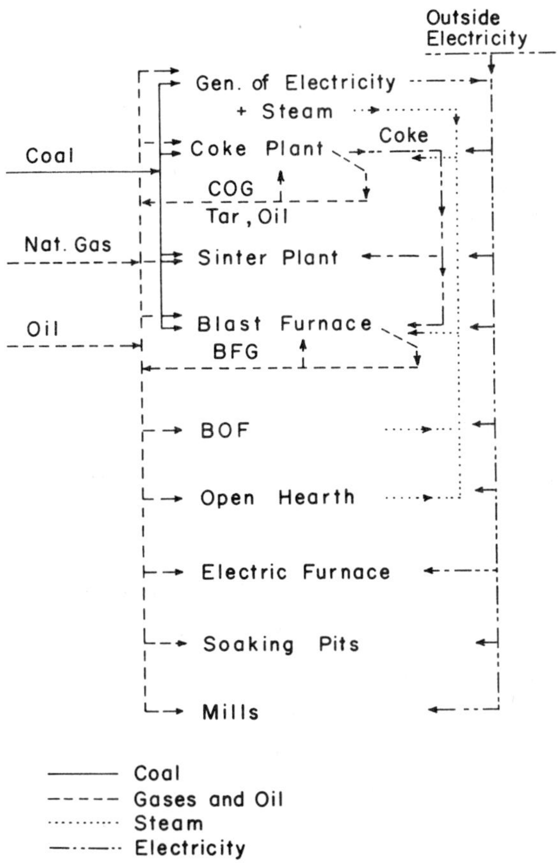

Fig. 1. Flow of fuels and electricity.

TABLE 2

Average Net Fuel Requirement for Agglomeration
(Basis: 1 Ton of Agglomerates)

Item	Amount (lb)	Energy/unit	Total Btu
INPUT			
Coke breeze	140	10,500 Btu/lb	1.47×10^6
Coal	20	13,500 Btu/lb	$.27 \times 10^6$
Gas for ignition	--	--	0.16×10^6
TOTAL			1.9×10^6
OUTPUT			
None	--	--	--
NET FUEL CONSUMPTION			1.9×10^6
ELECTRICAL ENERGY		42 kWh	0.42×10^6

The yield of metallurgical coke from coal of 69% by weight that was assumed in preparing Table 3 is typical of recent production statistics (1) for the American steel industry. A production of 15,000 cf of coke oven gas (COG) per ton of coke is taken to be about average, and the use of about one-third the thermal equivalent of that gas is required for underfiring. The difference between the fuel value of the coal processed and that of all of the product streams is charged against the process. All of the output materials, except for the small amounts of chemicals, can be used as fuels elsewhere in the plant. These fuels have a combined value of 8.5×10^6 Btu per ton of coke and constitute an important source of energy in an integrated steel-plant.

TABLE 3

Net Fuel Requirement for Coking
(Basis: 1 Ton of Coke)

Item	Amount	Energy/unit	Total Btu
INPUT			
Coal	1.45 ton	13,500 Btu/lb	39.1×10^6
OUTPUT[a]			
Coke	1 ton	12,500 Btu/lb	25×10^6
Coke breeze	174 lb	10,500 Btu/lb	1.3×10^6
Oil and tar	11 gal	155,000 Btu/gal	1.7×10^6
Gas (net)	10,000 cf	550 Btu/ft^3	5.5×10^6
Chemicals	30 lb	--	--
TOTAL OUTPUT			33.5×10^6
NET FUEL CONSUMPTION			5.6×10^6
ELECTRICAL ENERGY		30 kWh	0.33×10^6

[a]Total fuels other than blast furnace coke: 8.5×10^6 Btu.

The coke requirement of 1100 lb in the production of one ton of hot metal is consistent with present good American practice using a good quality burden and high-temperature air blast (2) as shown in Table 4. A modest amount of fuel injected in the tuyeres is assumed. While various fuels have been used for this purpose recently, it seems certain that the industry will move rapidly to adopt the use of coal or coal-oil slurry. With the stoves being about 50% efficient, a total of 4×10^6 Btu of fuel is assumed to be needed for heating the air. About 90% of this heat is obtained from blast furnace gas, but a small amount of

TABLE 4

Net Fuel Requirement for Producing Hot Metal from Agglomerates
(Basis: 1 Ton of Hot Metal)

Item	Amount (tons)	Energy/unit	Total Btu
INPUT			
Coke	0.55	12,500 Btu/lb	13.8×10^6
Fuel to tuyers	--	--	1.3×10^6
Net fuel to stoves	--	--	0.3×10^6
Steam for air	--	--	1.5×10^6
TOTAL			16.9×10^6
OUTPUT			
Net blast furnace gas[a]	--	90 Btu/ft^3	3.0×10^6
NET FUEL CONSUMPTION			13.9×10^6
ELECTRICAL ENERGY		12 kWh	0.12×10^6

[a]Plus the equivalent of 3.7×10^6 Btu blast furnace gas. Total blast furnace gas: 6.7×10^6 Btu.

high-quality fuel is also needed to obtain satisfactory combustion. Forty-five percent of the blast furnace gas is available for use elsewhere. The net fuel energy to produce 1 ton of hot metal is about 14 million Btu.

The fuel requirements for the production of steel by the basic oxygen process (Table 5) arises from the need for fuel for auxiliary purposes around the shop. Included also is the energy necessary for calcination of the flux, production of a nominal amount of ferroalloys, and generation of oxygen. The net fuel

TABLE 5

Net Fuel Requirements for BOF Operation
(Basis: 1 Ton of Raw Steel)

Item	Amount	Energy/unit	Total Btu
INPUT			
Misc. heating fuel	--	--	0.20×10^6
Burnt lime	150 lb	2,500 Btu/lb	0.38×10^6
Ferroalloys	7.5 lb	13,700 Btu/lb	0.10×10^6
OUTPUT			
Heat recovered from waste gases (steam)	--	--	(0.5×10^6)
NET FUEL			0.18×10^6
ELECTRICAL ENERGY			
General uses		26 kWh	
Oxygen generation	1600 ft^3 1.3 kWh/100 ft^3	21 kWh	
TOTAL ELECTRICAL ENERGY		47 kWh	0.47×10^{6a}

[a]10,000 Btu fuel/kWh.

requirement for the process is determined principally by what is done to recover the chemical and sensible heat in the waste gases. Some recovery of heat in the form of steam is usual; however, up to 0.4 million Btu of energy can be recovered by the use of a special collecting system for these gases.

Fuel requirements for basic open hearth operations using 40% scrap and a high consumption of oxygen are shown in Table 6. This is approximately average operations in an integrated steel plant in the past few years (1,2).

TABLE 6

Fuel Requirement for Basic Open Hearth Operations
(Basis: 1 Ton of Raw Steel, 40% Scrap, 60% HM)

Item	Amount	Energy/unit	Total Btu
INPUT			
Heating fuel	20 gal oil	150,000 Btu/gal	3.0×10^6
Burnt lime	20 lb	2,500 Btu/lb	0.05×10^6
Ferroalloys	11 lb	13,700 Btu/lb	0.15×10^6
OUTPUT			
Steam	--	--	0.20×10^6
NET FUEL			3.0×10^6
ELECTRICAL ENERGY			
Furnace operation	--	30 kWh	
Oxygen	1000 1.3 kWh/ 100 cf	13 kWh	
TOTAL		43 kWh	0.43×10^6

The energy requirements for the production of carbon steel on a large scale in the arc furnace are shown in Table 7. An item is included to cover the energy necessary to produce the coke and electrodes used in the process. The electrical requirements for melting, general uses, and production of oxygen are shown at the bottom of the table.

It is difficult to determine the energy required in heating and rolling because very little is reported on the matter, and because the amounts used vary greatly depending on the products made and the facilities employed. Statistics for 1972 show that

TABLE 7

Fuel Requirement for Electric Furnace Operations
(Basis: 1 Ton of Raw Steel)

Item	Amount	Energy/unit		Total Btu
INPUT				
Misc. heating fuel	--		--	0.20×10^6
Burnt lime	90 lb		2,500 Btu/lb	0.23×10^6
Coke and electrodes	30 lb		15,400 Btu/lb[a]	0.46×10^6
Ferroalloys	7.5 lb		13,700 Btu/lb	0.10×10^6
OUTPUT				
NET FUEL				0.99×10^6
ELECTRICAL ENERGY				
Furnace operations	--		500 kWh	
General uses	--		27 kWh	
Oxygen	200 cf	1.3 kWh/100 ft^3	3 kWh	
STEELMAKING TOTAL			530 kWh	5.3×10^6
Electrodes	10 lb	1.5 kWh/lb	15 kWh	
TOTAL			545 kWh	5.5×10^6

[a]Heating value and fuel for coking and graphitizing.

a total of 7.2×10^{14} Btu of fuels of all kinds were used for heating, reheating, and annealing by the industry. This is an average of 7.8 million Btu per ton of finished steel. (See Table 8.)

TABLE 8

Consumption of Fuels and Electrical Energy in Heating and Rolling Steel (Basis: 1 Ton of Finished Steel)

Item	Fuel (Btu)	Electricity (kWh)
All heating and rolling	7.8×10^6	410
Thermal energy for electricity	5.5×10^6	
TOTAL THERMAL ENERGY	13.3×10^6	
Soaking pits and primary mills	0.8×10^6	40
All mill operations other than item B	7.0×10^6	370

Analysis of the consumption of electrical energy by various processes and of the total consumption by the American industry indicates that about 30% of the total is used in the mills. For 1972, the analysis indicates that the consumption is approximately 410 kWh per ton of finished steel in rolling and finishing operations. On the assumption that 10% of all fuels and energy for rolling and finishing is used in the soaking pits and in the primary rolling operations, these steps require an average total of 0.8 million Btu and 40 kWh per ton of finished product. The total thermal requirement is 1.3 million Btu per ton of steel. The totals then for all other processing and rolling are 7 million Btu of fuel and 370 kWh of electricity. The total thermal requirement is 10.7 million Btu per ton of finished steel.

Of the reported (1) consumption of energy by the industry, approximately 10% of the total cannot be accounted for by accumulating the usages in the various segments of the processing system. A portion of the unidentified amount may be fuels of various kinds that are sold. Thus the accountable apparent use by the industry for the production of iron and steel in 1972 is approximately 31.0×10^{14} Btu. This is an average apparent use of 33.8×10^{6} Btu per ton of finished steel produced. As noted earlier this figure also does not represent the total energy required to produce a ton of finished steel.

COMPARISON OF PROCESSES

It is interesting to compare the overall energy requirements for the production of steel by the blast furnace-BOF sequence to those for the blast furnace-OH sequence and the melting of scrap by the arc furnace. The yields on the metallics charged to the steelmaking furnaces differ sufficiently so that it is best to make the comparison based on the production of finished steel. The requirements shown in Tables 9 through 11 are based on information contained in Tables 2 through 7. The yield from ingots to finished steel is the same for all three cases and is assumed to be 69% (1). It is seen that the total energy required through ingot production is essentially the same at 26 million Btu per ton of finished steel for the blast furnace-BOF and blast furnace-open hearth sequences. As is to be expected, the total energy required through steelmaking for arc furnace melting of scrap is much lower at 9.5 million Btu per ton of finished steel. The difference is 16.5 million Btu. These values apply for the production of steel starting with iron oxide concentrates.

The energy requirements with possible increased use of the QBM--Q-BOP type furnace in integrated steel plants should not differ materially from that shown for the BF-BOF sequence here. The effects of the slight reduction in use of hot metal and a slight increase in yield will probably be offset by the fuel required in the process. In the analysis of the requirements for scrap and hot metal, the BF--Q-BOP sequence is also lumped in

TABLE 9

Energy Requirements for Ingot Production for One Ton of Finished Steel by Blast Furnace-BOF (Basis: 1 Ton of Finished Steel) (1.45 tons of ingots; 69% yield, ingot to finished steel)

Steps	Amount (tons)	Energy (Btu)	Electricity (kWh)
Steelmaking	1.45 ingots	0.26×10^6	68
	0.5 scrap	--	--
	0.435 home	--	--
	0.065 purchased	--	--
Blast furnace	1.17 hot metal	16.3×10^6	14
Agglomeration	1.95 agglomerates	3.7×10^6	82
Coking	0.64 coke	3.6×10^6	19
TOTALS		23.9×10^6	183
FUEL EQ. OF ELECTRICITY		1.8×10^6	
TOTAL FUEL ENERGY		25.9×10^{6b}	

[a]Credit for melting scrap, 0.54×10^6 Btu for 0.057 ton finished steel, based on Table 11.

[b]Total fuel energy/ton of raw steel: 17.8×10^6 Btu.

with the BF-BOF sequence for two reasons: (1) because the relatively small decrease in the use of hot metal and increase in the use of scrap by the newer process should not alter significantly the projections for the BF-BOF sequence, (2) because it does not appear likely that the domestic steel industry will change over in a major way from BOF to Q-BOP operations within the next decade.

The final figures given in Table 9 and 10 are not suitable for judging the actual energy required for producing steel in

TABLE 10

Energy Requirements for Ingot Production for One Ton of Finished Steel by Blast Furnace-Open Hearth (Basis: 1 Ton of Finished Steel) (1.45 tons of ingots; 69% yield, ingot to finished steel)

Steps	Amount (tons)	Net fuel energy (Btu)	Electricity (kWh)
Steelmaking	1.45 ingots	4.35×10^6	43
	0.67 scrap	--	--
	0.67 home	--	--
	0.245 purchased	--	--
Blast furnace	1.01 hot metal	14.0×10^6	12
Agglomeration	1.68 agglomerates	3.2×10^6	71
Coking	0.56 coke	3.1×10^6	17
TOTAL		24.7×10^6	143
Fuel eq. of electricity		1.4×10^6	
TOTAL FUEL ENERGY		26.1×10^6 [a]	

[a]Total fuel energy/ton of raw steel: 18.0×10^6 Btu.

Note: Credit for melting scrap, 2.0×10^6 Btu for 0.21 ton finished steel, based on Table 11.

these two sequences. The reason for this is that there is a different mix of hot metal and scrap in each case. Inclusion of recycled home scrap in the charge is appropriate in any steelmaking operation because the practice is intrinsic to economical commercial operations. On the other hand, more purchased scrap can be processed per unit of hot metal in the open hearth than is possible in the BOF; hence, a correction for this effect is necessary. There are several possible methods for making this correction; the most reasonable appears to be to apply a credit

TABLE 11

Energy Requirements for Ingot Production
for One Ton of Finished Steel by Electric Furnace
(Basis: 1 Ton of Finished Steel)
(1.45 tons of ingots; 69% yield, ingot to finished steel)

Steps	Amount (tons)	Net fuel Energy (Btu)	Electricity (kWh)
Steelmaking	1.45 ingots	1.44×10^6	790
	1.58 scrap	--	--
	0.435 home	--	--
	1.14 purchased	--	--
Fuel eq. of electricity		7.90×10^6	
TOTAL FUEL ENERGY		9.34×10^{6}[a]	

[a]Total fuel energy/ton of raw steel: 6.55×10^6 Btu.

for the extra scrap (purchased) that can be handled. The credit is based on the best melting operation practiced commercially, i.e., arc furnace melting of scrap. Accordingly from Table 11, 1.14 tons of purchased scrap are required to produce 1 ton of finished steel with a consumption of 9.3 million Btu of energy through ingot production. Using the same figures for yield in the blast furnace-BOF sequence, the 0.065 tons of purchased scrap would produce 0.057 tons of finished steel and the prorated credit is 0.54×10^6 Btu. Similarly for the blast furnace-open hearth sequence the 0.245 tons of purchased scrap would produce 0.21 tons of finished steel and the credit is 2.04×10^6 Btu.

The comparison is slightly different if strand casting is employed instead of ingot casting. In this case the yield from raw steel to finished steel would be approximately 79% instead of the value of 69% as used in the previous discussion. The return scrap is approximately 20% and 30% of the raw steel, respectively, in the two cases. The calculation has been made in the same manner as was discussed previously, and the overall

TABLE 12

Effect of Casting Method on Total Thermal Requirements For Steelmaking to Slabs (Basis: 1 Ton of Finished Steel)

Item	BF-BOF		BF-OH		EF	
	Fuel 10^6 Btu	El (kWh)	Fuel 10^6 Btu	El (kWh)	Fuel 10^6 Btu	kWh
Ingot casting (69% yield[a])	23.9	183	24.7	162	1.4	790
added for primary mills	0.8	55	0.8	55	0.8	40
TOTAL	24.7	238	25.5	217	2.2	830
TOTAL THERMAL	27.1		27.7		10.7	
TOTAL ADJUSTED FOR SCRAP	26.6		25.7			
Continuous casting	21.3	163	21.6	143	1.3	690
(79% yield[b])						
TOTAL THERMAL	22.9		23.0		8.2	
TOTAL ADJUSTED FOR SCRAP	21.5		20.5		8.2	

[a]Yield of finished steel from raw steel.

[b]Energy requirement per ton of slabs can be obtained by multiplying each total figure by 0.79.

results are shown in Table 12. The first conclusion shown is that the total thermal energy required in any sequence is reduced markedly as the yield of finished steel from raw steel is improved. Thus, there is a strong motivation toward the use of continuous casting when energy is to be conserved.

With the use of strand casting with a yield of 79% of finished steel from raw steel, the total thermal requirements for making slabs from scrap by the arc furnace is 8.3 million Btu per ton of finished steel. The requirement for the blast furnace-BOF sequence is 21.5 million Btu and for the blast furnace-open hearth sequence it is 20.5 million Btu. Thus, the total thermal requirements for producing slabs from scrap is about 12 million Btu per ton of finished steel less than it is for producing slabs from hot metal. Based on 1 ton of slabs, the difference is 9.5 million Btu.

OVERALL ENERGY REQUIREMENTS

The total energy needed to produce one million tons of finished steel by each of several sequences of operations is shown in Table 13. These results have been obtained by using the information presented previously, but the energy required for making electrodes and ferroalloys has been omitted. Also shown are the amounts of scrap in addition to the available revert scrap that must by obtained from outside the steel plant in accordance with the need for raw materials set forth in Tables 9 through 12. Again the advantage of lower consumption of energy rests with greater use of scrap. Perhaps a better approach to the matter is that reduced use of hot metal reduces consumption of energy, but a related increase in the use of scrap from outside the steel plant is necessary to provide sufficient metallics for the required production of steel.

The advantage of strand casting in reducing the total energy requirements as shown in Table 13 will be realized if the ratios

TABLE 13

Comparisons of Energy Required to Produce 1 Million Tons of Finished Steel by Various Processing Sequences

PROCESS Sequence	BF-BOF	BF-OH	EF
INGOT CASTING			
Coal, 10^{12} Btu	25.2	21.9	--
Other fuels(net), 10^{12} Btu	6.1	10.4	8.4
Electrical energy[a], 10^{12} Btu	5.9	5.7	11.8
TOTALS(10^{12} Btu)	37.2	38.0	20.2
Purchased scrap[b], 10^{6} tons	0.065	0.245	1.14
STRAND CASTING			
Sequence			
Coal, 10^{12} Btu	22.0	19.2	--
Other fuels(net), 10^{12} Btu	5.5	9.3	7.5
Electrical energy[a], 10^{12} Btu	5.3	5.1	10.4
TOTALS(10^{12} Btu)	32.8	33.6	17.9
Purchased scrap[b], 10^{6} tons	0.19	0.34	1.13

[a] 10,000 Btu of fuel energy/kWh.
[b] 29% scrap in charge of BOF and 40% charge of OH assumed in each case.

of hot metal to scrap charged to the steel furnaces are maintained as shown in Tables 9 and 10. There would be an attendant reduction in the amount of energy required per unit of steel produced in the BF-BOF and BF-OH sequences, principally because the proportion of hot metal is reduced. Again there is a related increase in the need for scrap from outside the steel plant.

It is often argued that increased use of strand casting *reduces* the amount of scrap that must be obtained from outside the steel plant (3). An example where this would be true would be the case of an integrated steel plant which replaced ingot casting with strand casting and held the production of hot metal and finished steel constant. Because of the increased yield from raw to finished steel, less raw steel would be needed, the ratio of scrap to hot metal in the charge to the steel furnaces would fall, and the steel plant could become self-sufficient in scrap. In this case there would be only a small reduction in the amount of energy used per unit of finished steel produced only as a result of the lower production of raw steel and the elimination of soaking pits and primary mill operations. It can be argued that this condition might not pertain for long in a modern American steel plant as there would be great pressure to operate the steel furnaces at as high an output as possible, to increase the throughput of steel in the mills, and to avoid construction of expensive additional blast furnace capacity. Accordingly, the trend would be to operate the system in the manner outlined in Tables 9 to 13. One of the results then would also be a reduction in the energy required per unit of steel produced as the use of strand casting increases.

Recent experience with direct reduction indicates that approximately 16 million Btu are required to produce 1 ton of metallized ore. Unfortunately, this energy has to be supplied in the form of high-quality gas for the processes that have proven to be technically feasible. Hence, in the United States such processes are not practical for the near future because of the energy shortage. However, if one of these processes were used in conjunction with strand casting, the total energy requirement per ton of finished steel would be in the range of 57 million Btu.

Two cases that indicate the possible range of energy required in the production of steel in 1983 are shown in Table 14. The

TABLE 14

Effect of Strand Casting on Estimated Total Energy Requirement for Steel Production in 1983

2-1/2% Growth in consumption of finished steel
15% Imports of finished steel
Open hearth production is zero in 1983

Finished steel consumption 137.3 million tons
Finished steel production 116.7 million tons
Ratio of raw steel production BOF/EF 3.1

Process requirements	All steel ingot cast	All steel strand cast
Total raw steel: 10^6 tons	169.2	148.21
Energy needs: 10^{14} Btu		
Coal	22.2	19.4
Other fuels	7.8	7.0
Electrical energy	8.6	7.6
TOTALS	38.6	34.0
Metallics required: 10^6 tons		
Hot metal	103.2	91.8
Purchased scrap	38.2	49.0
Net Metallics	141.4	140.8

cases are for all steel being ingot cast, a condition closely approximated by the U.S. steel industry in 1972, or all steel being strand cast. The primary assumptions in the calculation are that assumptions in the calculation are that the growth rate in the consumption of finished steel will be 2-1/2%, imports of steel will be 15% of domestic production of finished steel, all open hearth operations will cease by 1983 and that the production of raw steel by BOF and EF operations will be in the ratio of 3.1/1, a ratio that prevailed in 1972. With a common production of 116.7

million tons of finished steel, no use of strand casting requires the production of 169.2 million tons of raw steel and 103.2 million tons of hot metal for steelmaking with the total consumption of 38.6 x 10^{14} million Btu of energy. It would be necessary to obtain 37 million tons of scrap iron outside the steel plant. In addition, it appears that new facilities to increase the production of hot metal (or other types of metallics from ore) by 7 to 8 million tons over present blast furnace capacity would have to be built.

With large-scale use of strand casting, the additional production of finished steel anticipated to be needed in 1983 over that produced in 1972 (approximately a 25% increase) could be achieved with an increase of total energy used directly in the production of steel of about 10%. No new blast furnace capacity would be needed, but the industry would have to obtain approximately 18 million tons of scrap from outside sources over that which it obtained in 1972. The two cases cited here are overly simplified to show that the energy needs of the steel industy are closely related to such matters as blast furnace capacity and the supplies of industrial and obsolete scrap. A more complete study of the requirements of the industry is in progress and will be reported later (4).

DISCUSSION AND CONCLUDING REMARKS

The analysis presented here shows that the energy consumed in producing hot metal is the main factor in determining the total energy required by the U.S. steel industry. Were metallized ore to be produced, the reduction step would also be the chief consumer of energy in the steel production flow sheet. The industry relies on coal as its principal source of energy, and it can be argued with considerable conviction that this is advantageous now to the industry because of the coal resources of the United States.

The date presented in Table 14 show that the dependence of the industry on coal relative to other fossil fuels may be increased in the next 10 years. This would be true particularly if the required electricity were generated with the use of coal and nuclear energy. It would also appear that the production of by-product oil, gas, tar, and pitch in the coking of coals is highly advantageous in times of energy shortage. This is in sharp contrast to the value placed on these by-products in the very recent past.

There are several pressures on the industry in the near term that can be expected to arise from the present energy shortage.

1. There will be an increased incentive to install strand casting and to take other steps to increase the yield of finished steel products from raw steel.
2. The industry will tend to use even larger quantities of scrap than it does now and there will be a general tendency towards a very tight scrap market in the years ahead.
3. In spite of the tight scrap supply to be expected, there will be a tendency to favor electric furnace operations over the BF-BOF sequence.
4. There will be opposing forces on blast furnace operations. Lower coke rates and the use of injected fuels will be used to increase the output of a given furnace. On the other hand, the value of by-products from coking operations as fuels in the steel plant will tend to favor the production of coke so that optimal plant operations may not be obtained with the lowest coke rates in the blast furnace.
5. The production of metallized ores in locations where gaseous fuels are cheap will pose an economic challenge to the domestic industry.
6. Commensurate with the long-term policy of the integrated steel producers of having multiple sources of iron ores to protect their supplies of raw materials, these producers will also have to continue to rely on blast furnaces located within the United States as the primary source of virgin metallics. It would appear that a number of new furnaces will be needed within the next 6 to 10 years.
7. Careful and economical uses of fuels at all stages of steel manufacture will be practiced. Recovery of heat frmm waste gases will be emphasized and it probably will be desirable to adjust blast furnace operations to obtain higher quality blast furnace gas which in turn can be used more widely in the steel plant.

8. The demand for steel is expected to continue to be strong because of the energy shortage. The reduced demand for steel because automobiles are smaller and lighter will be more than offset by the need for steel to build new plants for processing and producing fuels and for generating electricity. Plastics will be in a less competitive position than heretofore because of the very rapid rise in the cost of chemical feedstocks.

In the longer term, the industry must face competition from the transportation sector of the economy and the chemical industry for all kinds of liquid fuels including oils and tars that are products of its own coking operations. It may also find that coal itself will be preempted because of the specialized needs of these portions of the economy. The matter may be decided simply in the marketplace and by society's priorities on services and materials. Under such circumstances the industry will probably be faced with the use of process heat from nuclear sources, electricity from nuclear sources, carbonaceous residues obtained from fuel processing such as coal gasification and coal liquefaction, and the use of recycled and repocessed carbonaceous gases. The prospects of the industry participating in a "hydrogen economy" seem remote because of the premium that other sectors of the economy might be willing to pay for hydrogen as an energy resource. To meet these changes in its energy base, major alterations can be expected in the technology by which steel is made.

ACKNOWLEDGMENTS

The author is indebted to his colleagues at MIT and to many people in the metals industry for discussions with them on various aspects of the use of energy in the society and in the steel industry. Of particular value have been personal communications with members of the staff of the Battelle Memorial Institute at Columbus, Ohio.

REFERENCES

1. Annual Statistical Reports, American Iron and Steel Institute, Washington, D.C., 1960-1972.
2. Staff, U.S. Bureau of Mines, Minerals Industry Surveys, U.S. Dept. of Interior, 1960-1973.
3. C. E. Sims and A. O. Hoffman, "The Future of Electric Furnace Steelmaking," Proceedings, Electric Furnace Conference, AIME, vol. 30, 1092.
4. J. F. Elliott and F. E. Brantley, "Future Raw Materials Needs of the American Steel Industry." Presented at the Annual Meeting of AIME, Dallas, Texas, February 1974.

SOME IMPLICATIONS OF NATIONAL ENERGY PROBLEMS ON FUTURE STEEL INDUSTRY PROCESSES AND FUEL SOURCES

R. W. Bouman

Section Manager
Homer Research Laboratories
Bethlehem Steel Corporation
Bethlehem, Pennsylvania

INTRODUCTION

The steel industry in this country faces many serious problems in the future, and almost all of these problems are affected by the availability of environmentally acceptable fuels and reductants. For one, the steep rise in the cost of coal and coke will continue to have an appreciable effect on the cost-price squeeze. For another, the production and use of fuels in steel plants can cause environmental problems, and the elimination of these problems is increasing both operating costs and steel plant investment requirements. For these and other reasons, it would seem prudent for the steel industry to plan its fuel requirements systematically and with considerable foresight. Unfortunately, since the steel industry has little control over the factors affecting the energy dilemma in this country, planning for future energy supplies is risky and frustrating.

Because of the relevance of the steel industry's future energy supplies, much has been written and said on the subject. The general tone of these statements is not optimistic and can be briefly summarized as follows:

1. With decreasing availability of natural gas for the steel industry, gas will have to be replaced in the near future by

fuel oil, and possibly in the long run by coal. This situation is compounded by the fact that other industries and the electrical utilities are now facing the same problem, and this is resulting in a shortage of oil, particularly low sulfur heavy oil and distillate oil.

2. The steel industry will have to continue to rely on coal as a major source of energy for the rest of this century. While this projection is reasonable from a process viewpoint, it does present availability and cost problems because of mine safety legislation and concern for the environment.
3. The steel industry should improve process efficiencies and reduce energy requirements as much as possible.

While these general statements seem logical, they are not necessarily useful statements for determing industry strategy in energy matters. The objective of this paper is to present some insight into the steel industry's energy problems and possible alternatives to solve these problems. First, however, it will be helpful to review the energy dilemma faced by the entire nation.

THE NATIONAL PICTURE

The projected shortage of domestic energy supplies has been caused by a rapid increase in energy demand and by increasing resistance to the exploration, production, transporation, and use of fuels. The conflicts that presently exist among regulatory agencies, environmental and conservation groups, energy producers, and energy consumers have caused many delays in exploration programs and plant construction. Even though some of these problems will be resolved in some reasonable period of time, many will not be settled, and energy shortfalls will become more apparent in the next few years.

In order to discuss the energy shortfall meaningfully, a realistic projection of the total energy demand and available supplies is needed. There is no lack of energy projections, and it appears that an energy projection is available for anyone wishing to emphasize some special interest. The National Petroleum Council

(NPC) has made a detailed and well documented study (1) of energy supply and demand, and this study will be the source of much of the quantitative information presented below. This study has been widely read and used, but more importantly, it presents a broad spectrum of possibilities for future energy supply and demand.

In 1971 this country used about 69 quadrillion (10^{15}) Btu of energy (2), and, with a growth rate slightly less than that experienced in the 1960-1970 period, the projected demand for energy is about 102 quadrillion Btu in 1980 and 125 quadrillion Btu in 1985. The range of projected demands developed by NPC is from 96 to 105 quadrillion Btu for 1980, and 112 to 130 quadrillion Btu

TABLE 1

Projected United States Energy Demand
(quadrillion Btu)

Sector	Actual[a] 1971	Projected[b] 1980	Projected[b] 1985
Residential/commercial	16.3	22.4	26.6
Industrial	19.8	26.8	30.9
Transportation	17.0	23.9	28.3
Electricity[c]	11.9	22.8	30.2
Non-fuel	4.0	6.7	8.9
TOTAL	69.0	102.6	124.9
Range of Projections			
High	--	105.3	130.0
Low	--	95.7	112.5

[a]U.S. Department of Interior.

[b]National Petroleum Council, December 1972.

[c]Conversion losses only, electricity consumed is shown as part of other sectors.

for 1985, as shown in Table 1. These projections were developed by considering a range of possibilities for factors such as the cost of energy, population, economic activity, and environmental controls. Still other factors affect energy consumption in this country, and some prognosticators have gone so far as to emphasize other factors and project a decrease in energy consumption in the near future. While a decrease in energy consumption would be welcomed by most, it seems highly unlikely, and the NPC projections appear realistic.

Some items of interest in Table 1 for the steel industry are the amount of energy consumed by the industrial sector in 1971, the projected energy demand for the industrial sector in 1985, and the projected growth rate for the total energy demand. In 1971 the industrial sector consumed 20 quadrillion Btu, or about 30% of the total energy consumed. The projected industrial demand for 1985 is 31 quadrillion Btu of about 25% of the total demand. This projected decrease in the industrial share of the energy market is an indication of our increasingly service oriented economy and a shift in national priorities. It is not clear how this trend will affect heavy industry in times of fuel shortages.

The energy dilemma begins to become apparent when the projected demands as shown in Table 1 are compared with the projected availability of domestic energy supplies. As has been done with projected demand, a range of possible supply situations has been developed by NPC. This range of possibilities consists of:

CASE 1. High availability of domestic supplies due to a maximum development effort and a relaxation of current restrictions

CASE 2 AND 3. Intermediate availabilities of domestic supplies due to a relaxation in current development restrictions

CASE 4. Lowest availability of domestic supplies due to a continuation of the restrictions presently being imposed on energy developments

The projections for domestic supplies that would be available for each of these cases in 1980 and 1985 are shown in Table 2. The numbers in this table for each energy source are in the units normally associated with the particular energy supply as well as in quadrillion of Btu per year.

One qualification that should be made regarding the NPC resource projections in Table 2 concerns the expected nuclear generating capacity. NPC did not attempt to distinquish between coal and nuclear fuel for the electric utility sector, and this has resulted in projected nuclear capacities that are not realistic with the possible exception of Case 4. No criticism is attached to NPC's understandable decision to avoid the issue, but this does imply that either coal production will have to be higher than indicated in order to meet demands, or that the electrical energy shortfall will be very large. This is a serious problem, and it appears quite likely that an electrical energy shortage will occur before 1980.

A direct comparison of supplies in Table 2 to the projected intermediate demand in Table 1 shows a shortfall of domestic supplies in all instances except the 1985 supply situation for Case 1, i.e., where the development effort would be maximum and current restrictions would be realized. However, because complete interchangeability of fuels is not possible, a shortfall of domestic supplies has been projected by NPC for all combinations of supply and demand. These shortfalls, which translate into the need for imports, are shown in Table 3 as a percent of total energy demand. The implications of Table 3 are an increasing dependency on imported energy supplies and, as a result, an undesirable effect on our balance of payments. An important point to emphasize in Table 3 is that apparently little can be done with the need for large energy imports in 1975.

TABLE 2

Potential Domestic Energy Supplies for 1980 and 1985

Energy source	1971[a]	1980[b]				1985[b]			
	Actual	Case 1	Case 2	Case 3	Case 4	Case 1	Case 2	Case 2	Case 4
Oil[c]									
barrels (10^6)/day	11.3	13.8	13.0	11.7	8.9	16.9	14.4	12.3	10.5
Btu (10^{15})/year	22.6	28.2	26.6	24.0	18.1	34.7	29.4	25.3	21.6
Gas[c]									
cu ft (10^{12})/year	21.2	26.7	24.8	20.9	17.5	34.4	28.6	22.5	15.5
Btu (10^{15})/year	21.8	27.5	25.5	21.5	18.1	35.2	29.4	23.1	16.0
Coal									
tons (10^6)/year	511	851	734	734	705	1093	863	863	819
Btu (10^{15})/year	12.6	21.2	18.3	18.3	17.6	27.1	21.4	21.4	20.3
Nuclear									
MW capacity (10^3)	8.7[d]	188	188	150	107	450	375	300	240
Btu (10^{15})/year	0.4	11.3	11.3	9.8	6.8	29.8	25.2	20.2	16.1

Hydroelectric									
kWh (10^9)/year	266	296	296	296	296	316	316	316	316
Btu (10^{15})/year	2.8	3.2	3.2	3.2	3.2	3.3	3.3	3.3	3.3
Geothermal									
MW capacity (10^3)	-	10.3	5.3	4.5	2.5	19.0	9.0	7.0	3.5
Btu (10^{15})/year	-	0.8	0.4	0.3	0.2	1.4	0.7	0.5	0.2
TOTAL - Btu (10^{15})/year	60.2	92.2	85.3	77.1	64.0	131.5	109.4	93.8	77.5

[a] U.S. Department of Interior.

[b] National Petroleum Council, December 1973.

[c] Including synthetics.

[d] 14,700 MW as of December 1972.

TABLE 3

Percentage of Imported Energy Required to Meet Projected Demand (% of total demand)

NPC supply case	Actual 1971	1975	1980	1985
1	13	20	16	11
2	13	20	19	20
3	13	23	26	28
4	13	26	38	38
Projected demand, Btu (10^{15})/year	69.0	83.5	102.6	124.9

If this need for imported energy is satisfied with oil, Case 4, i.e., the worst domestic supply situation projected by NPC, would result in oil imports of 16.4 million barrels/day in 1980 and 19.2 million barrels/day in 1985. These possible import levels compare to oil imports of 3.9 million barrels/day in 1971. The effect of 19.2 million barrels/day of imported oil at $5 per barrel on the balance of payments is about $35 billion a year. In addition, even if a heavy dependence on imported oil were in the national interest, the effort required to site and build the needed deep water ports and refineries would be monumental, and with current restrictions would probably be impossible before 1980.

The range of possible domestic supplies that might be available is large, and in order to achieve a supply situation like NPC's Case 1, a number of national priorities must be established and carried out. These priorities include the following:

1. Increased exploration for oil and gas
2. More realistic procedures and policies for siting
3. More research funds to develop methods of converting high sulfur coal into a clean fuel

Some moves have been made in these directions; however, much more is needed if we are to achieve the degree of self-sufficiency indicated by NPC's Case 1. Unfortunately, many of the arguments being made against the development of domestic energy resources are not logical, and a clear course of action by legislators, regulatory agencies, and the courts is not apparent. To counter this state of confusion, there are proponents of a "national energy policy" that would resolve all or most energy conflicts. However, it is doubtful that this type of all-encompassing policy can be formulated and enforced in the near future. A realistic expectation for the availability of domestic energy supplies is in the range shown by NPC's Cases 2 and 3, and hopefully it will be closer to Case 2 than to Case 3.

The uncertainty presently associated with national energy matters will not be eliminated, and the steel industry, along with other industries, must make energy commitments for the future in spite of these uncertainties.

STEEL INDUSTRY ENERGY

In 1972, the steel industry in this country produced about 133 million tons of raw steel as reported by the AISI (3). According to my estimate, the energy consumed to make this steel was 3,647 trillion Btu; therefore, the energy consumed in the production and finishing of one ton of raw steel was about 27.4 million Btu. A breakdown of the steel industry's major energy for 1968 to 1972 is shown in Table 4, and a more complete review of energy used in the steel industry is shown in Table 5. Table 5 shows energy sources as normally purchased commodities; thus, oxygen, iron ore pellets, ferroalloys and lime are shown as separate entries even though some form of fossil fuel or electricity was used in their production. In addition, the electrical energy purchased directly by steel producers and the electrical energy used to produce oxygen and ferroalloys have been converted to resource energy units by

TABLE 4

Fuel and Electricity Consumption in the Steel Industry

Year	Coal [tons (10^6)]	Purchased electricity [kWh (10^9)]	Natural gas [cu ft (10^9)]	Fuel oil [gallons (10^9)]
1972	81.4	39.4	772	1.31
1971	79.4	37.5	603	1.11
1970	92.3	37.8	594	1.24
1969	90.9	36.7	634	1.24
1968	88.0	33.5	587	1.28

[a]Source: AISI Annual Statistical Report, 1972.

using 10,000 Btu/kWh in order to reflect power plant conversion losses. This method of presentation shows steel industry energy consumption to be slightly higher than has been reported previously. Other energy sources used directly by the industry like gasoline, LPG, and lubricating oils are small in comparison to the items listed in Table 5, and have not been included.

The most apparent trends since 1967 have been a decrease in the industry's dependence on coal and an increased dependence on natural gas. In 1968, coal accounted for 64.5% of the total energy used as compared with 58% in 1972. Natural gas consumption increased from 16.5 to over 21% in the same period of time. However, the Federal Power Commission, prompted by the natural gas reserve situation, has established a priority list of natural gas consumers. These priorities are based on end-use, with residential and small commercial consumers high on the list. It is apparent that large industrial consumers of natural gas, particularly those with boiler applications, will be low on the priority list. This means that the trend of increasing natural gas consumption for the steel industry since 1968 will be reversed.

TABLE 5

Energy Consumption in the Steel Industry
(trillion Btu and % of total)

Commodity	1972	1971	1970	1969	1968
Coal					
Amount	2116	2064	2400	2340	2288
%	58.0	61.4	64.4	63.4	64.5
Purchased electricity					
Amount	394	375	378	367	335
%	10.8	11.2	10.2	9.9	9.5
Natural gas					
Amount	772	603	594	634	587
%	21.2	17.9	16.0	17.2	16.5
Fuel oil					
Amount	183	155	174	174	179
%	5.0	4.6	4.7	4.7	5.0
Ferroalloys					
Amount	73	69	79	82	78
%	2.0	2.1	2.1	2.2	2.2
Oxygen					
Amount	25	22	24	23	20
%	0.7	0.7	0.6	0.6	0.6
Iron ore pellets					
Amount	41	35	37	35	31
%	1.1	1.0	1.0	1.0	0.9
Lime					
Amount	43	38	38	37	30
%	1.2	1.1	1.0	1.0	0.8
Total energy, Btu (10^{12})	3647	3361	3724	3692	3548
Raw steel, tons (10^6)	133.2	120.4	131.5	141.3	131.5
Btu (10^6)/ton raw steel	27.4	27.9	28.3	26.1	27.0

[a]conversion factors:

Coal	26 x 10^6 Btu/ton	Oxygen
Electricity	10,000 Btu/kWh	Iron ore pellets
Natural gas	1000 Btu/cu ft	Lime
Fuel oil	140,000 Btu/gallon	
	122 Btu/cu ft	
	600,000 Btu/ton	
	6.5 x 10^6 Btu/ton	

A direct result of the decreased availability of natural gas will be an increase in the use of heavy and distillate oil. This is simply a matter of expediency and it is doubtful that oil will be able to replace all the lost natural gas in the long term. Since the natural gas consumed in the steel industry is used largely for steel heating applications, the replacement fuel must be low in sulfur. A critical supply situation already exists for low sulfur heavy oil and distillate oil, and the long-range prospects for this supply situation are not good. In light of this situation, it is likely that the steel industry and other large industrial consumers of natural gas will eventually be using some form of converted coal instead of natural gas. The timing for this switch is very uncertain and is dependent on the establishment of national policies and priorities.

The federal government is now funding the development of a number of coal conversion processes, some of which could be applicable in steel plants. These conversion processes include high Btu gasification, low Btu gasification, solvent refining, and liquefaction. Whichever coal conversion process eventually becomes the most attractive, it will present a few additional problems for the steel industry. First and foremost, the converted fuel will be more costly than either natural gas or fuel oil. Published cost estimates for clean fuel from coal vary greatly, and the most recent estimates show that $1.50/million Btu, or higher, is quite probable.

In addition to cost, coal conversion plants can add still another environmental problem to steel plants. Coal will be converted into a clean fuel with process efficiencies of only 65 to 75%; thus the availability and treatment of cooling water can be a major factor in siting a conversion plant.

Another problem with clean fuel from coal is that large conversion plants will be needed to gain economy of scale. The clean fuel production from one coal conversion plant will be an order of magnitude greater than is needed by a large steel plant to replace natural gas. This will not be a major problem when the clean fuel from coal is marketable, or alternatively, is simply purchased from a utility company. However, since low Btu gas from coal is a good potential steel plant fuel that cannot be economically transported long distances, conversion plant size can be a problem for this energy source.

PROCESS CONSIDERATIONS

In planning for future steelmaking capacity, the industry will consider both operating and investment costs of the new facilities, and a significant portion of these costs is related to energy matters. Long-range planning of steelmaking facilities should consider the overall efficiency of the competing steelmaking schemes in addition to the availability and suitability of energy resources. For this purpose, a comparison of the energy requirements for two competing steelmaking schemes is of interest and will be presented.

The first scheme considered is the conventional steelmaking method which includes the coke oven, blast furnace, and BOF processes. In 1972 almost 75 million tons of raw steel, or 56% of the total, was produced by this method. A simplified energy flow diagram for the production of 1 ton of liquid steel in a BOF is shown in Fig. 1. It has been assumed for Fig. 1 that coal would be the only source of energy in the primary side of the steel plant and that all the coal tar produced in the coke ovens would be injected into the blast furnace. In addition, it was assumed that half of the virgin iron in the blast furnace would be from iron ore pellets and the other half would be in the form of sinter. This balance shows that about 15.4 million Btu are needed to produce 1 ton of liquid steel in a modern coke oven, blast furnace, and BOF plant. This energy demand cannot be compared directly to the historical performance in Table 5 since the 15.4 million

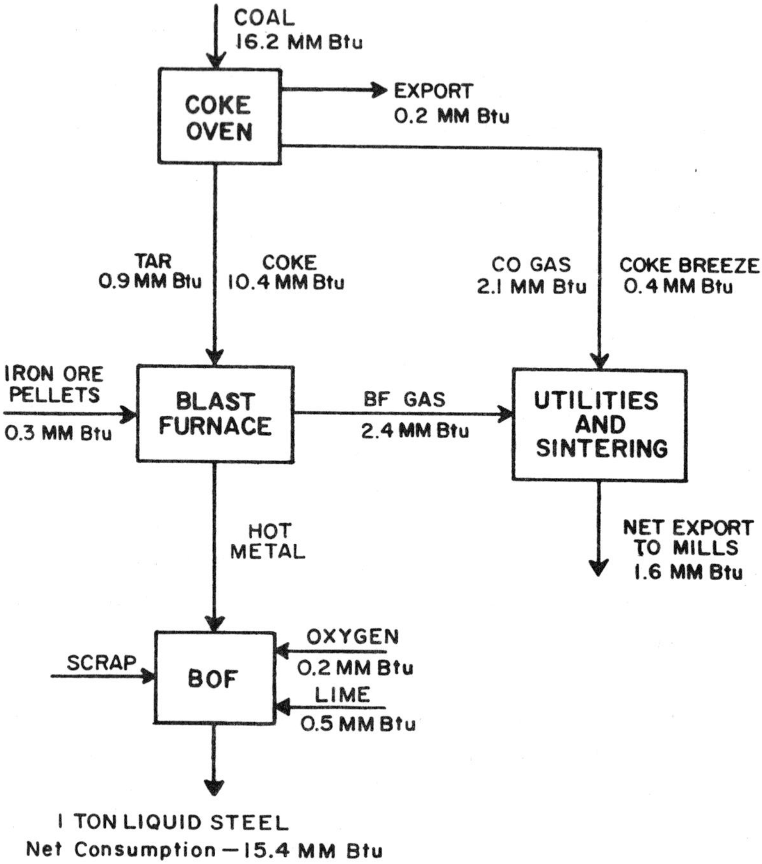

Fig. 1. Energy consumption to produce liquid steel in BOF.

Btu does not include the energy required to convert the liquid steel into finished products.

Another steelmaking scheme of interest is the production of sponge iron and the melting of this sponge in an electric furnace (EAF). Direct reduction (DR) processes using natural gas for the production of sponge iron have been proven commercially, and sufficient operating data are available (4,5) to estimate energy requirements. Assuming that only indurated pellets are used in the

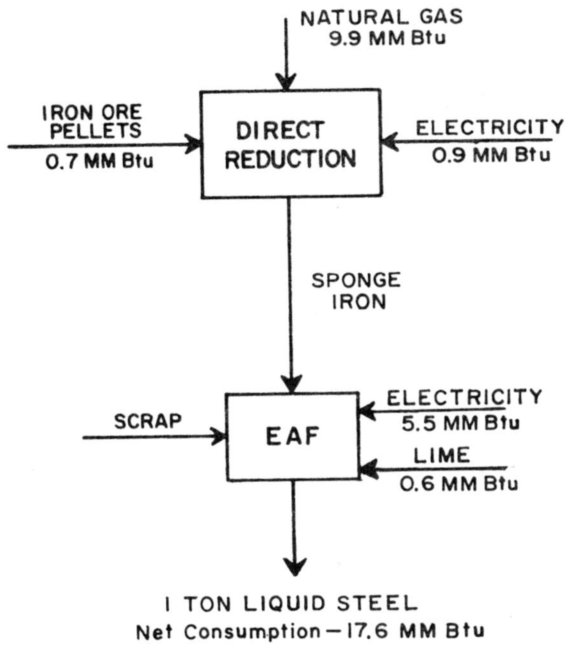

Fig. 2. Energy consumption to produce liquid steel with direct reduction and electric furnace.

reduction process and that the scrap charge in the EAF is the same as that for the BOF case, we arrive at the DR/EAF energy flow diagram shown in Fig. 2. This shows an energy requirement for 1 ton of liquid steel to be about 17.6 million Btu, or about 2.2 million Btu greater than for the BOF case. It is emphasized that all electrical energy requirements in Fig. 2 have been converted to energy resources by considering power plant conversion losses (i.e., 1 kWh is equal to 10,000 Btu).

The DR/EAF steelmaking scheme can be attractive because of low investment requirements, however the higher energy demand compared to the BOF scheme is not generally appreciated. This higher energy demand is not a significant problem where low-cost natural gas is plentiful, but this is certainly not the situation in the United States. Furthermore, the potential shortage of electrical generating capacity in the future casts additional doubts on the availability of energy needed for the DR/EAF scheme.

SUMMARY

This brief discussion of the projected energy shortage has pointed out that the steel industry will probably have to rely more on oil as a source of energy in the next few years and that in the longer run, the industry will have to use more coal. The comparison of steelmaking schemes is obviously limited in scope and not universally applicable, but it does point up part of the dilemma the steel industry faces because of the national energy situation. The steelmaking route that is the most efficient also requires higher investment costs. Similarly, the scheme that requires a least investment is also the least efficient and requires a premium fuel. Neither of these alternatives is particularly attractive for new steel plants in this country.

An alternative that is attractive from a process and energy viewpoint is the production of sponge iron in countries that have abundant natural gas supplies. This is not a new possibility and will become more attractive as the energy dilemma in this country becomes more pronounced.

REFERENCES

1. National Petroleum Council, "U.S. Energy Outlook," December 1972.
2. W. G. Dupree Jr., and J. A. West, "United States Energy Through the Year 2000," U.S. Department of Interior, December 1972.

3. Annual Statistical Report, AISI, 1972.

4. C. G. Sada, "The HyL Iron Ore Direct Reduction Process." Paper presented at IISI meeting, Johannesburg, South Africa, October 1972.

5. W. Korf, "The Midrex Direct Reduction Process." Paper presented at IISI meeting, Johannesburg, South Africa, October 1972.

ATOMIC ENERGY: ONE OF THE MAIN ENERGY SOURCES FOR STEEL-MAKING IN THE FUTURE

Werner Wenzel, Professor
Institut für Eisenhüttenwesen Technische
Hochschule, Aachen
German Federal Republic

For those who expect that heat from atomic reactors will be the solution of our present energy crisis, I must say at the beginning of this article that I have nothing to offer to today's trouble, but only for tomorrow's and in some cases even only for the day after tomorrow's.

It may be that the measures for developing this special branch of atomic techniques were not taken as fast as they could have been if we had forseen the energy crisis of these days, but no one will deny that nuclear energy takes front rank among those new energy sources that can be tapped by modern technology, due to the ready appreciation of the scope of its technical applications. I do not wish to forget the important role that could be played by the harnessing of solar energy, but I must restrict my remarks to nuclear energy, since only insufficient technical experience is available in this other field. We all know that the availability of nuclear energy in the future depends on certain technological requirements being met, the fulfillment of which is the subject of considerable activity.

I am assuming that, in principle, a solution will be reached, and that the problem of the fast breed reactor will have been solved, so that the easily accessible sources of uranium will not be rapidly exhausted. On this basis we can, in my opinion, assume that practically unlimited quantities of nuclear energy will be available for a very long time in the future, even if the technical realization of nuclear fusion does not appear to be attainable.

This optimism is based also on the fact that practically inexhaustible quantities of uranium are contained in seawater, which although present in only very small quantities, is nevertheless sufficient to secure supplies of uranium in times of scarcity. In the European nuclear research plant at Ispra, it has been calculated that in the coolant supplied to a seawater cooled nuclear reactor there is present a multiple of uranium content actually consumed by the reactor. I dare to be so optimistic as to prophesy that the engineers of the future will succeed in recovering and utilizing the required uranium from this coolant.

If, on the basis of the energy situation today, the iron and steel industry is to plan reasonably its development for tomorrow and the future, it can only do this after very careful consideration of the availability of the various energy carriers. The further one looks into the future, the greater is the importance assumed by nuclear energy as a source of energy in ironmaking.

FUNDAMENTAL POSSIBILITIES FOR THE APPLICATION OF NUCLEAR ENERGY TO IRONMAKING

The energy produced in nuclear power plants is available in two forms. Primarily, heat energy is produced in the nuclear reactor, which, in conventional power plants, is converted into electric power but which can, in principle, be made available as heat energy.

Both forms of energy differ widely in (1) the costs to the point of transfer to the consumer, (2) in the output efficiency measured against the quantity of energy produced in the nuclear reaction, i.e., the quantity of energy available to the consumer, and (3) as a consequence of these factors, the costs at the consumer end. Important differences arise also from the possible modes of utilization of the energy, in which the far greater scope of application must be ascribed to the electrical energy with its important advantage of transportability over large distances. Heat from nuclear reactors can probably be utilized only in the immediate neighborhood of a nuclear reactor and is limited to those consumers who can use the energy in the form of heat.

Possibilities for electrical nuclear energy in ironmaking

Since in the foreseeable future the largest proportion of nuclear power plants will pass on their energy in the form of electric power, the application of such energy to ironmaking merits special attention. The most important feature in assessing costs is the cost level attainable in the nuclear power plant. Ironmaking processes are already known in which electric power is used to deliver heat supplementary to the addition of fossil fuels as reductant. The electric low shaft furnace has, in recent years, been developed considerably further by prereduction of the ores with the aid of CO-rich exhaust gases from the main reduction plant. Today, current consumption figures of 1200 kWh/ton with an addition of 300 kg of reducing carbon per ton have been attained. However, the electric low shaft furnace is only viable for steel production if the costs for the electric power, as compared with the general cost level of the industrialized countries, are very low. We have to take into consideration that the electric low shaft furnace must be operated with nonbituminous coal, i.e., with coke, and that the coke consumption per ton of pig iron is only slightly lower than that in the most efficiently operated blast furnace. So a consumption of 1200 kWh electrical energy must be set against the 80 to 100 kg of oil consumed by the blast furnace in addition to the coke when taking into account the most modern blast furnace process, i.e., blowing in of reducing agents into the blast furnace shaft. It would seem that electrical energy from nuclear power plants could not be produced so cheaply as to give a positive balance to this cost relation.

A further important possibility to be investigated is smelting with electric power in the complete absence of fossil fuel as the reductant. This is the case where hydrogen is produced by the electrolysis of water and used to reduce the iron ore. The electrolysis of water has already reached a high technical level, especially as the result of developmental work in the United States. This has led to new processes with low energy consumption and relatively low investment costs due to high unit performance figures. On the basis of these figures we should consider how and under what conditions the electrolysis of hydrogen could be applied to ironmaking.

We have made a careful study of this in Aachen and have come to the conclusion that as regards both the production of hydrogen and the reduction of iron ores with this hydrogen there are no insoluble technical problems. It is mainly a question of developing sufficiently large units to ensure low capital costs. The schematic representation of such a plant is shown in Fig. 1.

A particular feature of this concept is that it offers the possibility of producing iron at any place on earth where iron ores are available. All that is required is a nuclear power plant and limited supplies of water. Furthermore, the reducing hydrogen could be generated from the condensed water recovered from the reducing exhaust gases.

Of course, the energy costs for the production of electrolytic hydrogen must be particularly low. With the present cost structure of the iron and steel industry, the cost of electrical energy must lie at the lower limit of the costs quoted for particularly large nuclear power plants. The cost structure is also dependent on whether we can succeed in utilizing the oxygen which is produced, such that this falls on the credit side of the balance. The quantities produced, however, are so great that up to now no complete application could be found.

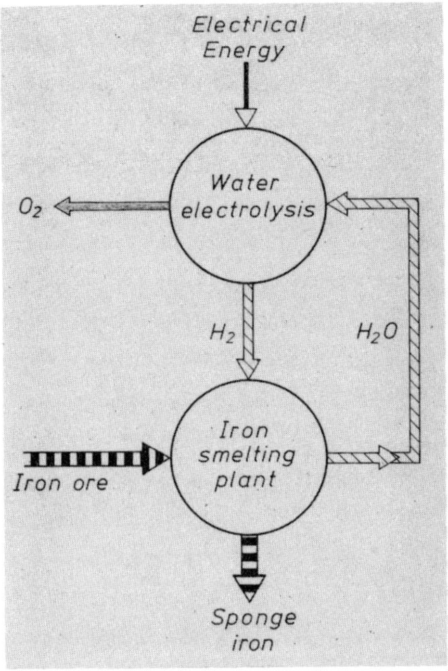

Fig. 1. Iron smelting with electrolytic hydrogen.

Ironmaking utilizing heat
from nuclear reactors

The production of iron from its ores requires a reducing agent and a heat supply for the reduction process and for the conversion of the crude iron into liquid steel. Reduction always takes place with gaseous reductants, so that the fossil fuels must first be converted into gaseous reductants, carbon monoxide and hydrogen. This requires heat at a temperature level necessary for the conversion of the fossil fuel. The lowest temperature level for such conversion at economic reaction rates lies at about 800°C. This temperature is valid, for example, for the conversion of brown coal. Other less reactive fuels require considerably higher temperatures which can exceed 1000°C as in the case of bituminous coal.

The heat from nuclear reactors must therefore be made available at correspondingly high temperatures, entailing the development of a reactor type capable of passing to the consumer a temperature of at least 1000°C or more, from its coolant. This leads to the development of high-temperature nuclear reactors operating with coolants that do not attack the graphite reactor material. Helium has proven to be the best coolant.

Quite independent of the problem of the direct application of nuclear heat, such high-temperature nuclear reactors have already been developed to produce electrical energy with the aid of the helium turbine. In principle, there are at present two types of reactors under development, the American Gulf and the British Dragon reactor, and the German pebble bed reactor. The Gulf reactor is suitable for coolant outlet temperatures of up to about 950°C. The pebble bed reactor, which at present operates at this upper temperature level, has the advantage that, in principle, temperatures of 1150 to 1200°C can be attained.

As will be seen later, the apparently fundamental limitation of the temperature of the Gulf-Dragon reactor to 950-1000°C must be considered as a disadvantage. The Gulf reactor, however, has the advantage that in its development it is at least three years ahead of its competitors.

The temperature advantage of the pebble bed reactor must, in my opinion, be reckoned very highly when considering the demands of iron ore reduction processes. The most economic reduction process is given when a reducing gas is generated from the conversion of fossil fuel with water and/or CO_2, which as regards both the outlet temperature and composition, can be fed directly into the reduction unit. All additional measures necessary to first cool such a reducing gas, change its composition, and finally bring it up to the required reduction temperature are very costly and affect the economics of the complete process. All our experimental work up to now has shown that it is difficult, if not impossible, to attain water contents in the final gas resulting from the conversion of natural gas, oil, or coal with steam at 900°C which is low enough to ensure sufficiently high reducing potential of the gas. Even if the reactor coolant were available at 950°C, 900°C would be the maximum upper limit of temperature available for the chemical reation. For this reason, supplementary electric heating has been considered for raising the peak conversion temperature up to 1000 or 1100°C. It is easy to understand that this also would bring complications. The right solution to the problem must lie in having reactor coolant available at a temperature of 1150 to 1200°C.

These considerations are especially valid for integrated plants in which the nuclear reactor and the iron smelting unit are installed as one complex. In this case the nuclear heat generated reducing gas can be fed directly into the reducing plant without loss of temperature. In the case of nonintegrated plants, i.e., where the gas generator and the iron reduction plant are set up at a distance from one another, the advantage of the high coolant temperature of the pebble bed reactor is not so marked. A cooling-off of the gas is necessary before its transport to the ironmaking plant, entailing a reheat for the reduction process. Other factors are the extent to which gas conversion processes can be coupled with the actual gas generation, and to which transport costs can vary with gas composition over large distances. The costs for piping hydrogen are less than for hydrogen and carbon monoxide mixtures or for relatively pure carbon monoxide. From this point of view there is much to be said for erecting centralized nuclear heat gas generators, producing hydrogen that can be transported over a distance to a reduction plant within an iron-and steelworks.

Besides production of reducing agents on the basis of fossil fuels with atomic energy there has been much talk in recent times of the fundamentally interesting application of nuclear heat for ironmaking: the thermal splitting of the water molecule to gaseous hydrogen and oxygen. This thermal splitting is normally possible only at very high temperatures and involves the problem of recovering separately the hydrogen and oxygen molecules.

Based on the work of De Beni and Marquetti in the European Nuclear Research Institute Ispra, many investigations are now taking place to attempt to break down this thermal splitting process into a series of successive chemical reactions, which, as a whole, operate with the intake of heat but which can be conducted at relatively low temperatures in experimental units already capable of construction. Hydrogen and oxygen are recovered as final products and the other reaction components are recycled.

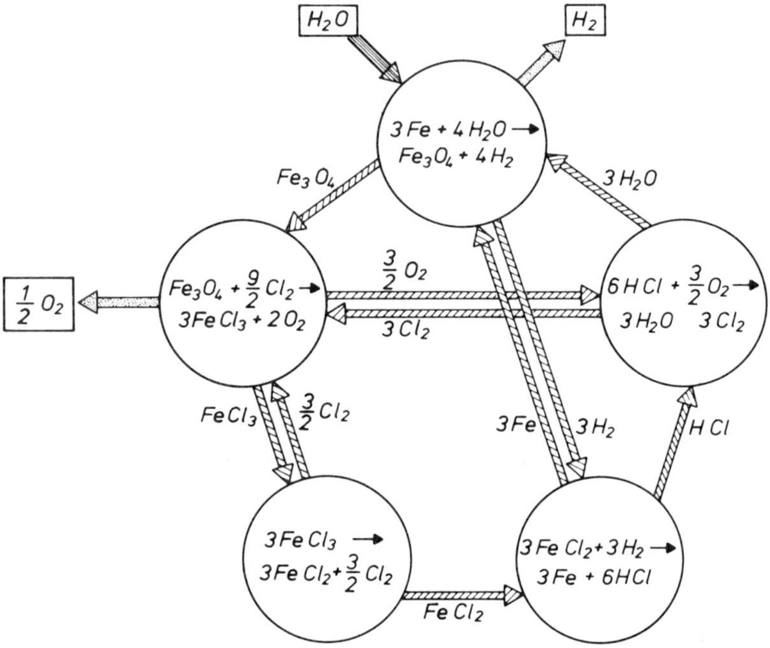

Fig. 2. The Knoche-process for thermal splitting of H_2O.

(1) $3Fe + 4H_2O \longrightarrow Fe_3O_4 + 4H_2$

(2) $Fe_3O_4 + \frac{9}{2}Cl_2 \longrightarrow 3FeCl_3 + 2O_2$

(3) $3FeCl_3 \longrightarrow 3FeCl_2 + \frac{3}{2}Cl_2$

(4) $3FeCl_2 + 3H_2 \longrightarrow 3Fe + 6HCl$

(5) $6HCl + \frac{3}{2}O_2 \longrightarrow 3H_2O + 3Cl_2$

Summation $H_2O \qquad H_2 + \frac{1}{2}O_2$

Fig. 2a. The Knoche-process for thermal splitting of H_2O.

The process system which I consider at present to be the most easily realizable--remembering that a large part of these processes operate under near utopian conditions--is that developed by my colleague Professor Knoche in Aachen, as shown in Fig. 2. Here we have in each case five single processes already well known in metallurgical and in chemical technology, but for which it is still necessary to evolve the optimum conditions for the application of nuclear based thermal energy at the various temperature levels. The thermodynamic feasibility of this process system has already been established. The total efficiency based on nuclear heat input at an upper temperature level of 800 to 900°C or at least 1000°C is about 70%, so that in principle conditions could be made favorable in the relatively near future, perhaps 10 to 15 years, for the production of hydrogen directly from water using nuclear heat.

The expert, however, will have no difficulty in detecting that the reactions used in the system of Professor Knoche are relatively complicated and costly, so that it is doubtful whether such a process could offer a better economic proposition than the electrolysis of water. My collegues and I are at present considering a simplified formulation of the main concept, in which an electrolysis takes the chief role, the products of electrolysis being converted through the input of thermal energy with auxiliary agents, so that hydrogen and oxygen are recovered as the final products with the auxiliary agents being recycled. In such a system two or three conversions are sufficient; two, in the limiting case. The advantage would lie in the relative simplicity of the processes. The basic idea is the heating of the auxiliary processes by means of the waste heat from the current generation for the electrolysis, e.g., from the waste heat of the helium turbine. Fig. 3 shows such a scheme.

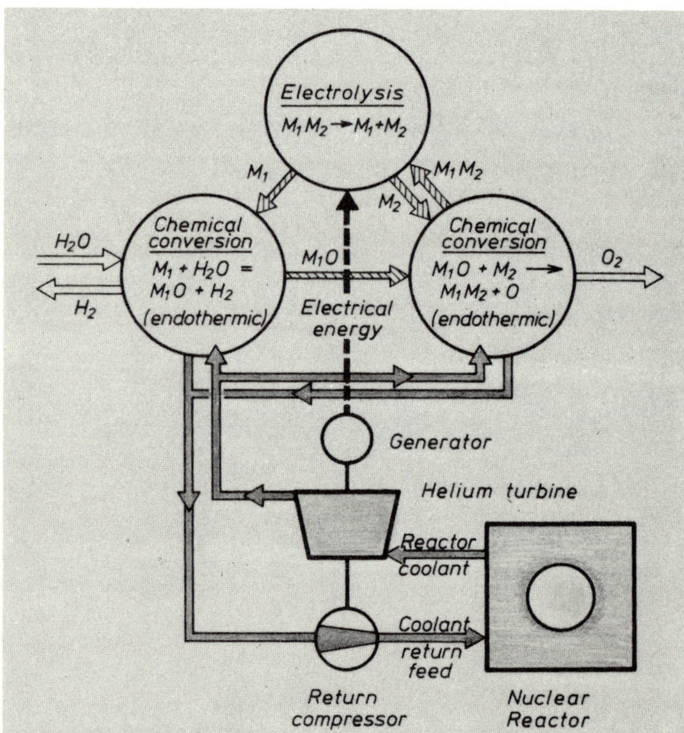

Fig. 3. Production of hydrogen with nuclear energy through combination of electrolysis and endothermic chemical conversion.

Summarizing, we can say that the application of nuclear energy will lead to the production of combustible gases either by electrolysis or by the conversion of fossil fuel consisting mainly of cabon monoxide and hydrogen, which will be distributed by means of large pipeline systems. Hydrogen will probably play a great role as pipeline gas. The various production plants would be coupled to this extensive hydrogen supply system and then operate on the basis of nuclear heat, according to the various process systems either mentioned here or yet to be developed. The iron and steel industry would also take from this supply system the required quantity of reducing hydrogen where a closed system is not already present, i.e., where nuclear plant with gas generator is combined with the iron-and steelworks.

ACTUAL PROCESSES FOR THE APPLICATION OF NUCLEAR HEAT TO IRONMAKING

When we started in the Aachen Iron and Steel Institute more than 10 years ago to develop iron ore reduction processes using thermal energy from nuclear reactors, we had first in mind a direct reduction process of iron oxide with carbon. A whole series of experments were undertaken, some under extreme conditions, such as the heating of ore-coal briquettes at very high pressures, in an attempt to realize this concept with the low temperatures available from the nuclear reactors at that time. Finally, however, we came to the conclusion that the thermodynamic, kinetic, and thermal problems to be overcome were so large that no success could be expected within the foreseeable future.

The much simpler way, the method being followed today, proved to be the conversion of the reducing agent by means of nuclear based thermal energy to a reducing gas based on hydrogen, mixtures of hydrogen and carbon monoxide, and less frequently, carbon monoxide. This reducing gas is then fed into the actual iron ore reduction process.

The advantage of using a two-stage process is that in the reduction stage processes can be utilized which have been developed in the last 10 years. Such processes are HyL, Midland-Ross Purofer, NU-Iron, etc., which I need not further elaborate. Also, as regarding the generation of reducing gases, known techniques in the petrochemical industry could be utilized, such as the steam reforming of methane, where the only change required would be the method of heating by helium or a secondary heat transfer medium in place of exhaust gas.

I do not, however, wish to minimize the considerable problems of the two stage process, particularly in the safety field, which arise from heating with nuclear energy and from coupling of a reduction plant with a gas generator and a nuclear reactor. The important question is whether it is possible to heat a methane reforming tube directly with reactor coolant helium. At all costs one must prevent the reactor coolant from entering the reduction gas, due to radiation components arising from the nuclear fuel. With direct heating of the methane reforming tube with reactor coolant, contamination of the reducing gas can occur due to leaks in the tubing system, to gas diffusion through the tube material, and to accidents involving breakage of the tubes.

To investigate these conditions, a single tube plant was built in the German Nuclear Research Unit in Jülich, in which a single methane reforming tube of technical dimensions was heated with helium. The experimental diagram is shown in Fig. 4. The results were so encouraging that in Jülich it was concluded that the contamination of the product gas through radiation components could be supressed to a very low level, even with this direct heating.

On the other hand, a number of workers in Germany and in other countries are tackling the problem using an intermediate heat transfer cycle in which the intermediate heat transfer medium may be helium, steam, nitorgen, or liquid lead. This solution

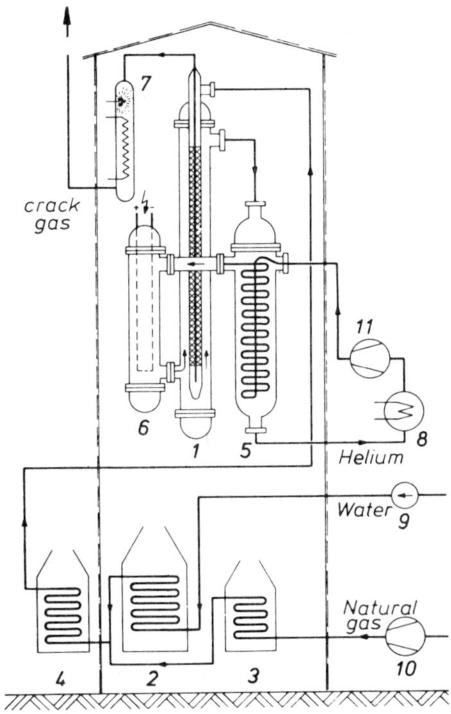

Fig. 4. Experimental setup in Jülich for gas reforming with (simulated) nuclear heat.

1. Cracking reactor
2. Steam raiser-superheater
3. Natural gas heater
4. Steam-natural gas superheater
5. Helium heat exchanger
6. Helium heater
7. Crack gas cooler
8. Heat exchanger
9. Feed water pump
10. Natural gas compressor
11. Helium circulatory blower

goes a long way to meeting the safety demands, but involves a considerable increase in investment costs and also a loss in temperature gradient of at least 50°C.

With direct heat transfer it is convenient to combine the steam-reformer, the steam generator and, if necessary, the process gas heater together with the reactor core in one unit, and to

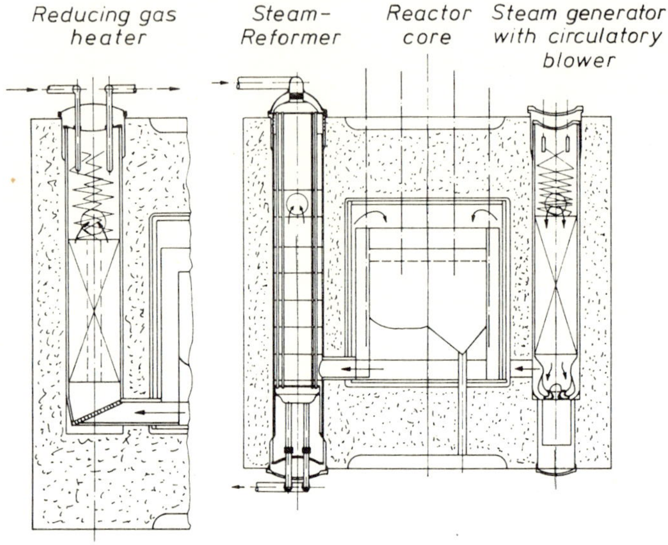

Fig. 5. Combined unit of atomic reactor with steam-reformer and gasheater in one pressure vessel.

house the whole complex in one reinforced concrete chamber. A schematic diagram of such a complex can be seen in Fig. 5.

Apart from the fundamental problem of the generation of reducing gases with nuclear based thermal energy, much work is going on to examine the various process possibilities of an integrated gas generator and ore reduction plant. The variations here are so many that I can only make reference to some of the important features.

In principle, there are two possibilities in applying reducing gas made with the aid of nuclear heat to ironmaking. The present trend is to produce sponge iron with this reducing gas and to melt this sponge iron in an electric furnace. Such developments aim to make plants largely self-sufficient, which means that the remnant waste heat in the reactor coolant after gas generation should suffice to produce electric power for the melting of the sponge iron,

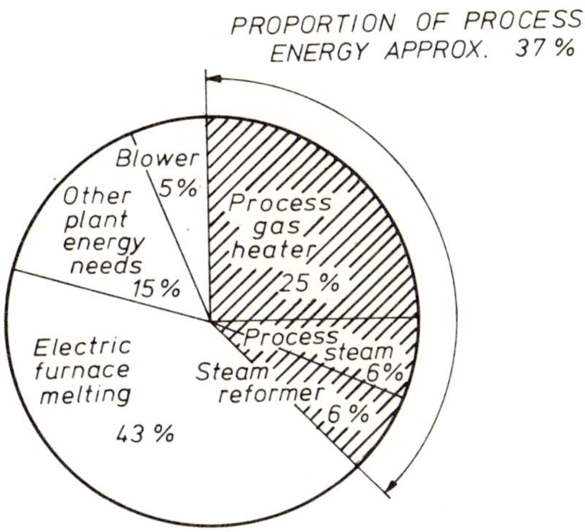

Fig. 6. Distribution of total energy from atomic reactor for an integrated steelworks. From "Kerntechnik," 1973, Nr. 1-3.

and to cover any other needs of the steelworks. This self-sufficient mode of operation is indeed possible, as can be seen from Fig. 6, which shows the distribution of the total energy input from a nuclear plant to the various take-off points in an integrated iron-and steelworks.

The mode of operation depends largely on the type of redution unit utilized. With present day techniques the shaft furnace is suitable only for relatively low reducing gas pressures. The fluidized bed, however, allows considerable pressures to be maintained without difficulty reaching in the neighborhood of 40 atmospheres in the reactor core. This makes the fluidized bed reactor for iron ore reduction especially interesting, since it allows the same pressure level to be maintained throughout the complete system of reactor core, generator, gas purifier, and reduction unit. In the interests of equal stress distribution at the various parts of the heat transfer surfaces, this is clearly desirable, and goes a long way to fulfilling the condition of operation without an intermediate heat transfer cycle. Such an integrated gas generator and reduction process system is shown in

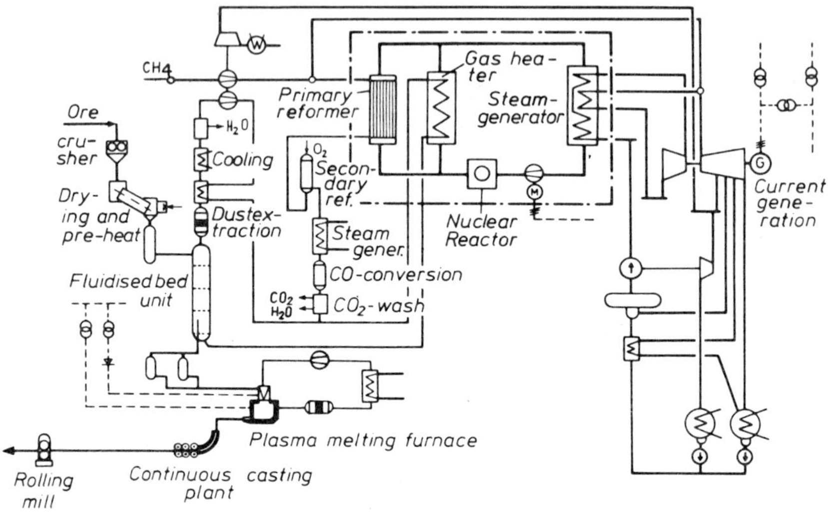

Fig. 7. Integrated gas generation with reduction process with atomic heat.

Fig. 7, which indicates a simplified development. As in the case of similar integrated systems developed by other working teams, the actual reducing gas is hydrogen, which is produced by the steam reforming of methane, with subsequent conversion of CO, and washing out of the CO_2 according to well-known methods. Hydrogen as reducing medium has a considerable advantage in that, after washing and condensing of the steam in the reducing exhaust gas, the remaining hydrogen can be recycled into the reducing gas after reheating. As against this, hydrogen reduction is an endothermic process, so that considerable quantities of heat must be supplied to the fluidized bed reduction. Fortunately reduction with hydrogen can be carried out with a sufficient reaction rate within a temperature range of 600 to 650°C, and the required temperature rise of about 200°C can be supplied by heating the reducing gas in the nuclear reactor to about 850 to 900°C.

The second application is closely related to well-known conventional methods. It would be of particular advantage if the blast furnace process could be altered with the aid of nuclear

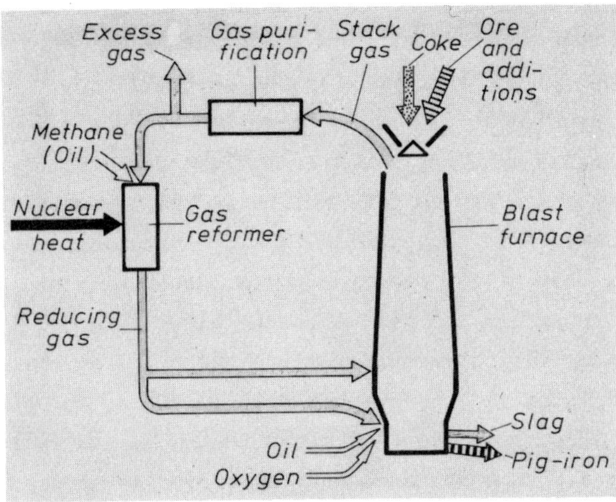

Fig. 8. Scheme for an integrated gas generation-reduction process with injection of reducing gas in the BF.

energy, such that the metallurgical coke requirements could be reduced. The production with nuclear heat of cheap reducing gases from cheap fossil fuels, such as natural gas and oil, leads naturally to various process possibilities for injecting this reducing gas into the blast furnace, resulting in saving coke.

In Aachen we have been engaged on the optimizing of a process, which aims at reducing coke consumption to a minimum and maintaining the total energy consumption at the lowest possible level. Fig. 8 shows schematically such an integrated gas generator reducing plant based on nuclear heat, in which the reducing gas is injected not only into the stack but also into the hearth of the blast furnace.

Optimum conditions for the heat consumption in the blast furnace process are attained if part of the top gas, rich in CO_2 and H_2O, is converted with a fossil fuel such as methane, to produce a reducing gas which can be blown into the stack, bringing about nearly 100% indirect reduction. The combustion energy of the coke in the hearth would then be used chiefly for melting the pig iron and slag, for reducing out the accompanying elements silicon, manganese, and phosphorus, and for the carburization of the iron.

One problem which has come to light with the injection of reducing gases into the stack of a blast furnace is that these side-blown gases are only effective in the boundary zones of the blast furnace. In our process this disadvantage is eliminated by injecting part of the reducing gas produced with nuclear heat, together with oxygen and oil, into the blast furnace. This reducing gas replaces the nitrogen normally introduced by the blast into the blast furnace. Even in the middle zone of the blast furnace, indirect reduction takes place and the whole gas cycle is relieved of nitrogen. Nuclear based thermal energy is applied in this process to the gas converter. Although no new problems arise, it may be that the gas conversion cannot be carried out within the pressure chamber of the reactor core, and that a secondary heat transfer cycle has to be led out of the reactor core and coupled to a gas conversion plant near the blast furnace. In any case it would be desirable to have the nuclear reactor in fairly close vicinity to the blast furnace.

It should be pointed out that such further developments in blast furnace operation would lead to a considerable increase in performance in terms of per cubic metre furnace volume or per square metre groundspace, due to the elimination of nitrogen from the hearth, and stack gas.

COST FORECAST AND PROBABLE FUTURE DEVELOPMENT

Although, as a result of the scarcity of fossil fuels, development of nuclear energy in ironmaking will one day lead to large-scale technical application, the actual timing of this will depend largely on the economics. For this reason, much attention has been paid to producing cost figures. In many cases, however,

ASSUMPTIONS	
Cost for	
Ore (67% Fe)	9.162 RE/ton
Natural gas	0.0136 RE/Nm^3
Coke	20.8 RE/ton
Nuclear heat	1.136 RE/million kcal
Electric power	0.0032 RE/kWh
Man power	2.080 RE/hr
Capital costs for	
Nuclear sponge iron route	4.628 RE/ton of liquid steel
Blast furnace/LD route	7.264 RE/ton of liquid steel
PRODUCTION COSTS	
1 ton of liquid steel by	
Nuclear sponge iron route	41.224 RE
Blast furnace /LD route	45.583 RE
1 RE = calculation unit = 3,6 DM = ≈1.5 U.S.-$ (1973)	

Fig. 9. Results of the cost analysis for the production of steel with the aid of nuclear heat as published by Euratom 1970.

simplicity and apparent workability of the various processes put forward have counted for more than the working out of detailed cost advantages, which often involve some degree of complication. One such cost study was worked out by the European Community Euratom during the years 1968 to 1969 and published in 1970. This study, in which I took part, is already four years old, particularly with respect to some important technical details and their cost basis. Nevertheless it gives important indications on the principal economic possibilities of the application of nuclear energy to ironmaking.

Figure 9 shows the most important assumptions and results of the Euratom study. The cost data are reduced to so-called calculation units, as is usual in the European Community. One calculation unit here represents about DM 3.60. It would at first appear to be difficult to relate these cost data to the American price level, since the cost assumptions refer to a location in West Europe, in particular to the erection of an iron- and steelworks on the coast. Also, in Europe, costs have changed considerably in the last four years, the cost of coke, for example, has nearly doubled since 1969,

From the Euratom study, however, it can be seen that the production costs for liquid steel using nuclear heat and methane as the energy basis are, at least, not higher than for conventional iron-and steelworks with the blast furnace and LD process. These relationships would also be valid for the United States, where both natural gas and coke are cheaper, so affecting both routes of manufacture.

This Euratom study has been checked in many places, including an enlarged working team of the European Community. The results of the enlarged study are not yet at hand, but it can already be established, that the final results will not differ greatly from the original Euratom study even if now more emphasis is placed on a comparison between the future nuclear energy ironmaking techniques and the likely future developments of the conventional techniques.

Under normal conditions the integrated nuclear plant iron- and steelworks will operate more cheaply than the nonintegrated type, where the separate energy producer and reduction plant are coupled by a gas pipeline. The cost advantage could be of the order of 5 to 10% of the final costs, depending on the location.

The prevailing energy situation and future cost estimations have encouraged considerable research activity aimed at making nuclear ironmaking a reality in the near future. In Germany this trend is already bound up with a certain tradition and the realization of these plans may certainly be counted upon, even if the setting up of a prototype iron- and steelworks does not seem probable before 1975, and the completion of a large iron- and steelworks not a commercial proposition before 1985.

Work on the realization of such a project appears to have gone furthest in Japan, where, according to my information, there is already a development team with considerable industrial scope cooperating with a committee of experts on the technical execution of such projects. The financing of such a prototype iron-and steelworks of the order of 7.7 billion Yen* appears to be secure. In England, also, recent intensive research activity should result in early realization, but in Italy, despite the existence of definite plans, lack of funds appears to be the stumbling block due to the general economic situation.

I shall not refer to the situation in the United States as I am sure there are people here at this conference who are more qualified to talk on this than I.

The new atomic era, which up to now has had little repercussion in the iron and steel industry, could receive a completely new impulse, if environmental awareness, i.e., the trend toward processes with minimum pollution, were to remain the determining factor in our industrial activity. Here, the application of nuclear energy would offer new possibilities of which ironmaking from water electrolysis offers the optimum solution, but where nuclear heat produced sponge iron, too, represents a considerable improvement over the present conventional iron and steel production via the blast furnace and LD converter. To the best of my knowledge these considerations have played a particularly big role in Japan in encouraging measures for the development of these new techniques.

---

*Equivalent to about 32 million U.S.-$ at 1973 exchange rates.

ENERGY CONSERVATION IN THE STEEL INDUSTRY

A. C. Elliott
Superintendent, Utilities Department

and

J. L. Uvira
Research Technical Adviser

The Steel Company of Canada Ltd.

The demand and supply situations for traditional fossil fuel energy are on a collision course in most of the highly industrialized countries. That sounds like a "beautiful" newspaper headline and, I suppose it is. However, we in the industrial sector of the economy don't regard it as beautiful. To us, as I'm sure we're all aware, it is the "beast, rather than the beauty". So much for my journalistic endeavors, now let's get down to business.

We are being continuously informed of possible energy shortages in the United States which will consequently affect Canada, since the economies of our countries are closely tied together. Canada is expected to supply a portion of the United States' desired energy (oil and natural gas) and we can therefore also expect intermittent local energy shortages, especially in Ontario, which imports about 80% of its fuel.

About 40% of Ontario's energy was supplied by oil in 1970. Natural gas accounted for close to 20% of which only a very small quantity came from local fields. Another 20% of the energy was generated from coal, mostly imported from the United States. About 18% of the province's energy was derived from water power. Give or take a percentage point or two, these represent the total sources of Ontario's energy. Since our company consumes a considerable portion (about 2.5%) of the fuels consumed in our province, we are devoting a great deal of effort to develop and improve methods to conserve energy in our plants.

Let me first outline the make-up of The Steel Company of Canada, Limited. We product approximately 40% of the nation's steel with our current capacity of 5,660 million ingot tons per year. This will increase to more than 6 million ingot tons when expansion programs, now underway, are completed. Our operations are fully integrated and our mines are located in both Canada and the United States. Our manufacturing facilities, comprising 18 plants, are situated in Ontario, Quebec, Alberta, and Saskatchewan. Our largest plant is Hilton Works in Hamilton, Ontario, and the energy conservation at that Works is the subject of my remarks today.

Prior to discussing our approach to the conservation of energy, I would like to outline some facts and figures concerning energy usage in our operations.

TABLE 1

Internal Energy Sources at Hilton Works
The Steel Company of Canada, Limited

Energy Sources	10^9 Btu
Total product fuels:	37.836
Blast furnace gas	(22.720)
Coke oven gas	(15.116)
Total waste heat	3.867
OH Furnaces	(2.800)
BOF	(0.356)
Reheating Furnaces	(0.711)
Total from Hilton Works Sources	41.703

What are our own internal sources of energy?

As Table 1 indicates, we derive a considerable quantity of chemical heat from our primary operations (blast furnaces and coke ovens), but substantially less (in the form of sensible heat) from steelmaking and rolling operations. We feel that these latter areas may offer us a real opportunity for capturing waste Btu's if we can find, adapt, or adopt the proper technology.

How much energy did we consume in our operations at Hilton Works during the past year and which type of fuel provided the lion's share of our Btu requirements?

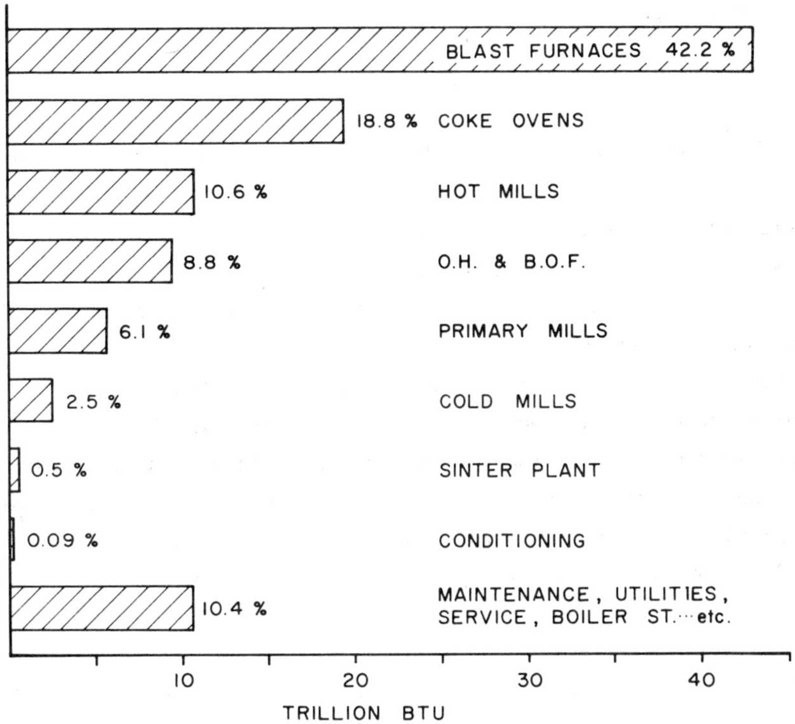

Fig. 1. Energy consumed by departments at Hilton Works in 1972. Data supplied by the Utilities and Electrical Departments.

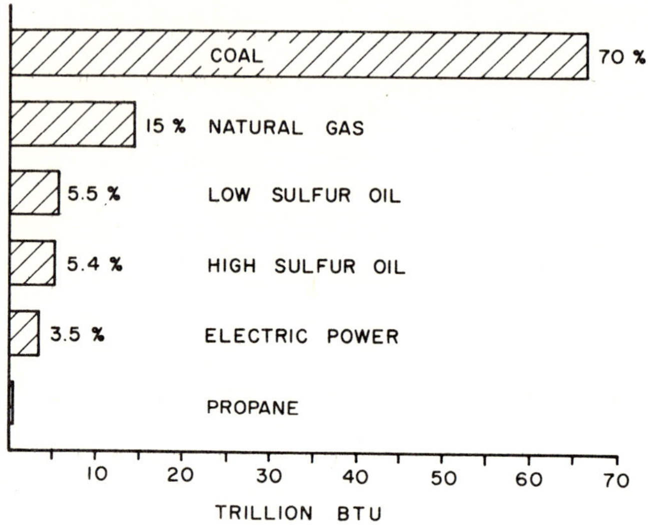

Fig. 2. Fuel and electric power consumed at Hilton Works in 1972. Data supplied by the Utilities and Electrical Departments.

Figures 1 and 2 illustrate what all steelmakers know, at least as far as trends go. The vast majority of our energy is consumed in the initial primary operations, namely the coke ovens and the blast furnaces. Any steel company, such as ours, is highly dependent upon coal (and the ensuing coke) for reducing the iron ore in the blast furnace and producing molten pig iron. In this respect, we are much different from the average large industrial complexes. "King Coal" still reigns supreme in our industry as it is presently operated. As Fig. 2 reveals that about 70 trillion Btu's, or about 70% of our total energy supply in 1972 came from coal.

How much did we pay (per million Btu's) for the different fuels?

In an energy intensive industry such as ours, costs of our Btu's are of the utmost importance. Figure 3 shows that electricity was our most expensive energy source in 1972, although we did not use

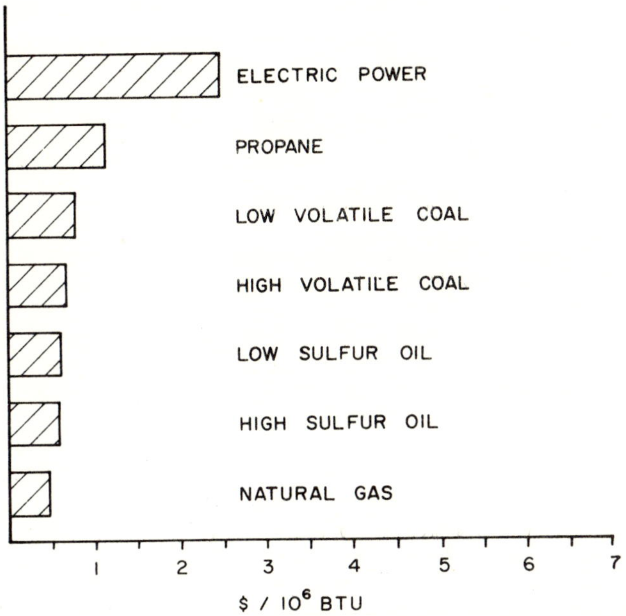

Fig. 3. Cost of purchased fuel and electric power at Hilton Works in 1972. Cost data supplied by the Accounting Department.

a proportionately large share for our operations as Fig. 2 indicated. I would like to point out, however, that even though this figure shows gas and oil as our cheapest energy source in 1972, the situation in 1973 and 1974 is extremely different, although we do not as yet have out final estimates and quotations for these periods. It may well be that we will have to turn more to coal useage or to electricity (generated at either coal fired or nuclear stations) if our industry is to remain commercially viable.

How much in fact, will energy cost us in the future? Figure 4 gives some idea of our expert's projections for the cost of energy in the future. We obviously have no precise "crystal ball," but as I noted a moment ago, energy in general is certainly

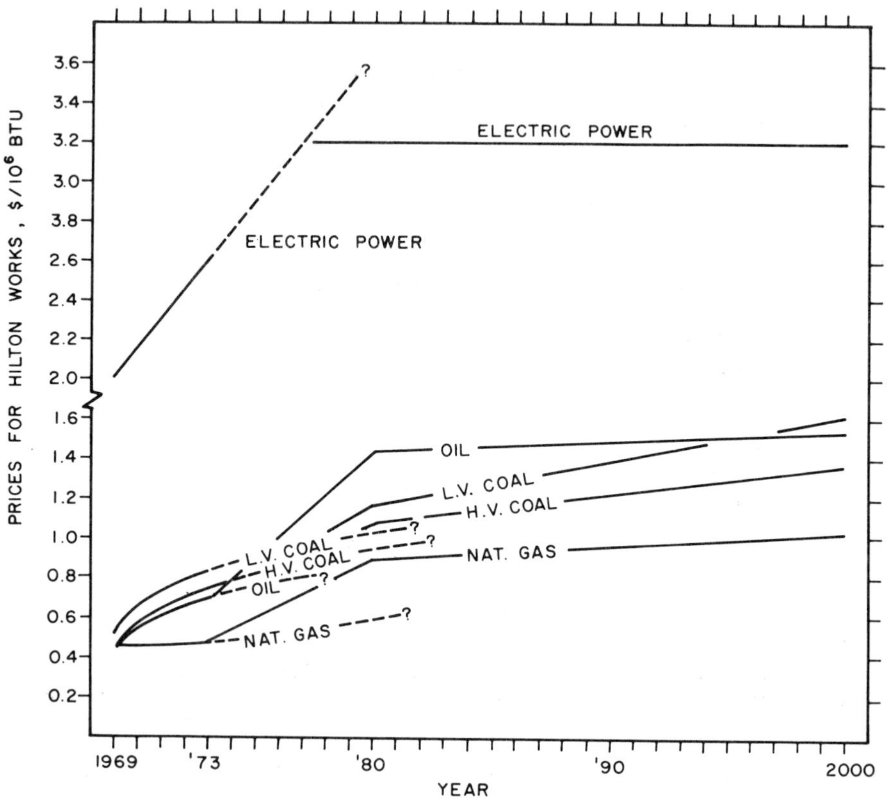

Fig. 4. Fuel prices in past five years and projected prices. Data for 1969-1973 supplied by the Accounting, Utilities and Electrical Departments.

going to become more expensive and definitely scarcer. We and, I am sure, other steel companies have come to the conclusion that steel made in the future must be produced using substantially less energy than that used in our present operations. This could mean total process changes and/or continued efforts to further harness waste heat lost from some of our units.

Let us look at some of the ways we might achieve our objective.

1. We must do away with the energy "joyride" in our plant, i.e., prevent overheating, overcooling, and overlighting our buildings, stop the idling runs, etc. Perhaps all of us, even as individuals at home, are guilty of these types of energy wasting activities.

2. We must increase the efficiency of energy conversion, since our power factor has been decreasing over the last four years and our load factors are still lower than 75%. We have also not yet exhausted our efforts to economize reheating and annealing of steel.

3. We have to take a closer look at our energy conservation. The bleeding of coke oven gas and blast furnace gas, for example, must be minimized. In the past year, we bled more than 200 million cu ft of coke oven gas and 2 billion cu ft of blast furnace gas, which, given the proper conditions, could have been used elsewhere in our operations. The numbers may be large, but the overall percentage is small, under 1%. Nevertheless, each percentage point counts!

4. We should attempt to utilize tar, the coke oven by-product which is produced in copious quantities, to the greatest extent possible as a source of internal Btu's. Previously, in attempting to sell the tar, we were faced with a buyer's market situation. Not anymore.

5. We must intensify our waste heat recovery program. Potential waste heat has yet to be recovered from blast stoves and from many of the existing reheating furnaces in out universal slabbing mill, new bloom and billet mill, and strip mill.

6. Finally, we must alter the existing metallurgical processes as much as we can and, in addition, select future processes consistent with the most abundant and most economic forms of energy.

These ideas sound rational and reasonable on paper. Their implementation, however, is something else again.

Let us now discuss the details involved in resolving some of these problems. A few months ago, we organized an "idea trigger session" of senior staff members. These men were selected from departments other than Utilities and Engineering, in order to obtain some ideas on conservation of energy at Stelco from people who might have a fresh look on this subject. We eventually collected 153 ideas which then were assessed by competent pesonnel, evaluated, and ranked. Some of them became our objectives which I have touched on above.

ELECTRICAL DEPARTMENT
ENERGY CONSERVATION REPORT

Mill____________ Report No.____________

Area____________ Elect File 0-L4-1 R&D File 15-36-41

Summary

KWH Savings/Year____________ Dollar Savings/Year____________

Estimated Cost of Recommended Changes____________

- -

CONSERVATION OF ENERGY - UTILITIES AND AUXILIARIES

CONTACT: Hans Kreplin, 2736

Unit or Process	Proposed Alteration	Nature of Saving	Remarks

Fig. 5. Forms used for data on energy saving.

As one of our first concrete moves, we launched an attack against the wastage of energy. Two engineers were temporarily released from their duties in the Utilities and Electrical Departments to (1) identify the areas of potential wastage of energy, (2) start organizing groups of volunteers who would set an example of how to use energy more economically, and (3) come up with ideas of how the user could be made more interested in saving energy.

Figure 5 is an example of the type of form which we have adopted in an attempt to aid company personnel in pinpointing and possibly resolving problem areas which lead to energy wastage.

In a very short time, we have uncovered many interesting situations. As examples, I would like to outline several areas where energy wastage has already been discovered and steps taken to remedy the problems.

1. One of our smaller buildings at one time was used as a stores area which required considerable candlepower for the employees of this department. After the building was renovated for office use, a portion of it was left unused, but the lighting intensity was unchanged. As a result of our survey, we have now greatly reduced the lighting requirements (~50% less) and thus, in this one small building alone, we have saved about $240 worth of electricity per year.

2. One of our larger office and service buildings used to be equally bright day and night. (In hindsight, it seems rather foolish to have lights glaring on unused desks and cabinets at night.) Now, when the office staff leaves at 5:00, the lights are automatically dimmed to a level suitable for building maintenance and cleanup. After this time, they are further reduced in intensity to a light which is just adequate for security purposes. By taking these steps, we have saved $72,000 worth of electricity per year. Needless to say, programmed lighting will be introduced into all of our office areas as soon as possible.

Along these lines, I might mention in passing that I have noticed some people wearing sweaters in our offices during the hot summer days. This is certainly food for thought.

3. We discovered that a turbine driven combustion air fan could have its static pressure control setting lowered from 9 to 7 WG (with still adequate performance). This move saved us about $5,000 per year.

4. One of our major discoveries and alterations dealt with a billet reheating furnace. In this particular furnace, we used a constant combustion air flow and modulated the fuel only. Originally, there were reasons for doing this (e.g., decarburization of the steel) but these conditions no longer apply. We have now gone to fuel/air ratio control and in the process saved ourselves some $100,000 worth of natural gas per year--not a small item.

These are but a few of the energy wastage areas which we have turned up so far. With the right frame of mind instilled in our employees and using appropriate financial awards, we are convinced that our energy wastage can be trimmed to the bone. One of our employees put it well, I think, when he noted that even a miniscule wastage of energy at the plant (with our enormous usage) can have a tremendous effect at *his* home in terms of *his* available energy, now or in the years to come.

This investigation is still proceeding and we expect that in the near future, more areas of potential energy savings will be uncovered in our plant.

Our next move will be to attempt an increase in the conservation of energy, which will result in a further increase in the harnessing of existing internal energy sources at Hilton Works as was already indicated in Table 1.

As everyone well knows, significant quantities of energy can be conserved in every metallurgical plant by introducing waste heat boilers, heat exchangers, etc. It is also generally known that, under present circumstances, the use of this equipment tends to decrease the average return on investment. As the fuel prices go up, however, the energy conservation aspect will become more attractive.

The energy supply, availability, and cost in the future presents a complex problem, which cannot be left only to departments such as Purchasing and Utilities to handle.

It will require a coordinated, company-wide effort incorporated into a broad far-reaching program. We have already initiated such a program with the definition of areas of interest and goals (Fig. 6). Our Energy Program is guided by the Research & Development Department, since we feel that a great deal of research work has to be done not only in the field of fuels but also in the area of metallurgical processes in order to most economically utilize the available energy.

I might point out here, that our Company, in conjunction with Atomic Energy of Canada, Ltd., has been vigorously pursuing the possibilities of utilizing nuclear energy in our operations. While it is unlikely that any commercially viable nuclear steelmaking process would evolve before the year 1990 or even 2000, we feel that we must attack the problem now in order to meet, or hopefully beat the deadline, at which time the traditional fossil fuels might become totally uneconomic for us in spite of our continuing efforts in the energy conservation field.

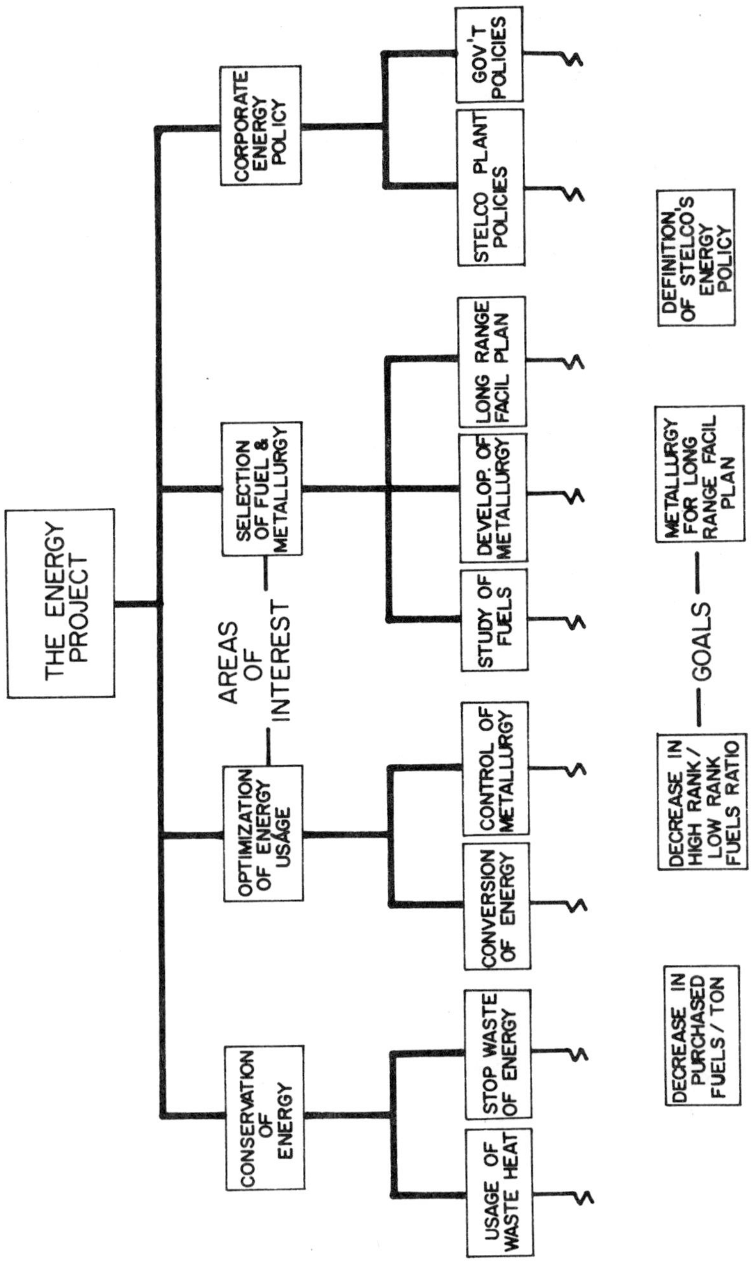

Fig. 6. Outline of the energy project at Stelco.

In closing, I would like to emphasize that the purpose of our short presentation is not only to indicate our activities in the energy field, but also, hopefully, to trigger a wide ranging discussion which may help the steel industry determine an appropriate course in securing energy for present and future operations.

ENERGY PROBLEMS AND MEASURES TO BE TAKEN AGAINST THEM IN THE JAPANESE STEEL INDUSTRY*

Mikio Takagi

General Manager
Energy Control Department
Nippon Steel Corporation

DEVELOPMENT AND RECENT CHANGES IN THE SITUATION OF THE JAPANESE STEEL INDUSTRY

Japan's postwar economy has achieved a remarkable development. In this favorable economic environment, the steel industry of this country recovered from its war devastation and attained the position of the world's largest steel supplier possessing strong international competitive power, with its annual crude steel production for 1970 reaching 92 million tons or 4.6 times that a decade before--a growth unprecendented throughout the world.

The steel industry has been enabled to contribute largely to the realization of the high-rate growth of the Japanese postwar economy chiefly by the following circumstances:

1. The domestic steel demand has expanded rapidly owing to both the high-rate growth of the national economy and the promotion of the heavy and chemical industrialization of Japanese industries.
2. The steel industry has succeeded in realizing remarkable productivity improvements with the positive investments of companies in technological innovations, such as the construction of large-scale coastal steelworks.
3. The steel industry has been able to enjoy low cost, stabilized supplies of iron ore, coal, and other raw materials necessary for the fast increasing steel production in spite of the fact

---

*Paper presented by Prof. Shintaro Tabata, Executive Director, Iron and Steel Institute of Japan, Tokyo, Japan.

that this industry must depend upon import for the most part of such raw materials.

4. With the cost reduction realized by mass production and equipment rationalization since the 1960's as a background, this industry has been able to expand export in coping with the increase of the world steel demand.

However, the goal of the high-rate growth policy of the Japanese economy has been clearly achieved, the fundamental direction of economic operation to be aimed at next is turning about from the heretofore pursued rapid growth, supported by private equipment investments and export promotion, to a new pattern of growth utilization oriented to "the perfection of national welfare" and "the promotion of international cooperation." Amidst this current of economy, large changes are now coming over into the steel industry also. The typical ones are environmental pollution problems and resources and energy problems, both of which are most closely related to energy.

Therefore, it is considered that the steel industry should make its utmost effort to develop new techniques to cope successfully with these problems by fully exploiting its brains, technology, and capital so that it can approach limitlessly near freedom from pollution and pave the way for an integrated resources and energy policy. It is imperative for this industry to fulfill its mission as a stabilized supplier of basic materials indispensable for the construction of a high-welfare society only after it has made such technological developments.

PRESENT CONDITIONS OF ENERGY CONSUMPTION IN THE JAPANESE STEEL INDUSTRY

Position of the steel industry as seen from the aspect of energy consumption

The rate of energy consumption of the steel industry to total energy consumption is as shown in Table 1. It is characteristic that, while the figures for the United States and Europe are 5.2%

TABLE 1

Rate (%) of Energy Consumption of the Steel Industry to Total National Energy Consumption[a]

	United States			Europe			Japan		
Energy breakdown	1960	1970	1980	1960	1970	1980	1960	1970	1980
Energy for steel industry	6.0	5.2	4.4	12.4	10.3	8.7	16.2	20.7	19.1
Kinds of energy									
Coal	20.8	30.3	41.8	13.4	24.9	41.7	16.4	68.2	83.0
Petroleum	-	0.7	0.6	4.1	3.7	3.3	10.9	7.8	9.0
Electric power	5.2	2.9	1.8	8.4	8.2	6.4	13.1	14.5	12.6

[a]Source: OECD Energy Commission

and 10.3% (1970), respectively, the figure for Japan is 20.7% which is considerably higher. When the energy consumption of the steel industry is observed according to kinds of energy, the figure for coal in the case of Japan is very high, near 70%, which is expected to make further increases. The fact that the rate of coal consumption is high is advantageous to Japan in the sense that in Japan, fuel lower in sulfur content than heavy oil is being used. The steel industry, especially one of an integrated iron- and steelmaking system, may be said to be the only industry using coal energy effectively.

The final energy consumption rates of the steel industry in Japan are shown in Table 2. The energy consumption of the steel industry accounts for 17.7% of the total national energy consumption. A breakdown of this by kinds of energy reveals that the coal group accounts for 80.1%, electric power 16%, and the petroleum group 6.8%. The component rates by kinds of energy are 54.1% for the coal group, 22.7% for the petroleum group, and 23.2% for electric power. The figure for the coal group is above 50%.

TABLE 2

Final Energy Consumption Rates
for the Steel Industry: 1971
(10^{10}kcal)

Type of energy	Steel industry only			Total for Japan		
	Quantity	Component rate A(%)	Component rate B(%)	Quantity	Component rate A(%)	Component rate B(%)
Electric power	12,260	23.2	16.0	76,687	25.8	100.0
Coal group	28,537	54.1	80.1	35,628	12.0	100.0
Coal	318	0.6	8.8	3,606	1.2	100.0
Coke	17,910	33.9	93.4	19,167	6.4	100.0
COG	4,175	7.9	89.0	4,690	1.6	100.0
BFG	6,134	11.6	86.7	7,072	2.4	100.0
Others	--	--	--	1,092	3.7	100.0
Petroleum group	11,969	22.7	6.8	176,137	59.2	100.0
Gasoline	--	--	--	19,681	6.6	100.0
Naphtha	--	--	--	23,481	7.9	100.0
Kerosene	--	--	--	26,239	8.8	100.0
Heavy oil	11,969	22.7	14.6	82,096	24.2	100.0
Others	--	--	--	24,639	8.3	100.0
Natural gas	2	--	--	1,986	0.7	100.0
Town gas	--	--	--	5,470	1.8	100.0
Others	--	--	--	1,477	0.5	100.0
Total	52,768	100.0	17.7	297,385	100.0	100.0

NOTE: Component rate A is the rate of consumption of each kind of energy to the total energy consumption of the steel industry. Component rate B is the rate of energy consumption of the steel industry to the total consumption of each kind of energy.

TABLE 3

Demands of Primary Energy & By-product Gas Energy in the Japanese Industry[a]

Item	Unit	1965	1966
Production of crude steel	10^3T	41,296	51,898
Production of pig iron	10^3T	27,930 0.668	33,534 0.670
Apparent consumption of crude steel	10^3T	28,375	39,504
Amount of export (crude steel base)	10^3T	31.4 12,952	24.1 12,514

Item of Energy		Demand	% in Calorie	Demand	% in Calorie
Primary Energy					
Coking coal	10^3T	14,050	41.0	16,024	40.2
Coal for general use	10^3T	590	1.3	568	1.1
Anthracite	10^3T	176	0.5	173	0.4
Purchased coke	10^3T	6,885	19.3	8,361	20.1
TOTAL		(159,424)	62.1	(184,714)	61.8
Petroleum coke	10^3T	45	0.1	43	0.1
Heavy oil	13^3kl	4,895	18.8	5,566	18.4
Electric power	10^6 kWh	17,436	16.6	20,930	17.2
Purchased oxygen	10^6M^3	400	0.3	561	0.3
Purchased coke oven gas	10^6M^3	1,313	2.1	1,576	2.2
GRAND TOTAL		(257,443)	100.0	(299,359)	100.0
Energy					
Blast furnace gas	10^6M^3	41,655	14.6	52,432	14.0
Coke oven gas	10^6M^3	4,478	7.3	4,883	7.8
BOF gas	10^6M^3	300	0.2	420	0.3
TOTAL		(56,898)	22.1	(66,224)	22.1
Primary energy/ton of crude steel	10kcal	6,234		5,768	

[a]Source: Official Statistics of the International Trade and Industry.

TABLE 3 (continued)

Item	Unit	1967	1968
Production of crude steel	10^3T	63,777	68,987
Production of pig iron	10^3T	41,581 0.645	47,849 0.693
Apparent consumption of crude steel	10^3T	51,362	50,965
Amount of export (crude steel base)	10^3T	20.1 12,836	26.3 18,161

Item of Energy		1967 Demand	1967 % in Calorie	1968 Demand	1968 % in Calorie
Primary Energy					
Coking coal	10^3T	18,959	39.0	21,287	39.0
Coal for general use	10^3T	340	0.5	297	0.4
Anthracite	10^3T	184	0.3	195	0.3
Purchased coke	10^3T	11,011	21.7	12,628	22.2
TOTAL		(224,630)	61.5	(253,574)	61.9
Petroleum coke	10^3T	123	0.3	230	0.5
Heavy oil	10^3kl	26,621	18.0	7,054	17.0
Electric power	10^6kWh	25,122	16.9	28,011	16.7
Purchased oxygen	10^6M^3	961	0.4	1,567	0.6
Purchased coke oven gas	10^6M^3	2,224	2.9	2,784	3.3
GRAND TOTAL		(365,447)	100.0	(410,405)	100.0
Energy					
Blast furnace gas	10^6M^3	61,125	13.4	67,303	13.1
Coke oven gas	10^6M^3	5,975	7.9	6,784	7.9
BOF gas	10^6M^3	544	0.3	722	0.4
TOTAL		(78,668)	21.6	(87,849)	21.4
Primary energy/ton of crude steel	10kcal	5,730		5,949	

TABLE 3 (continued)

1969			1970			1971		
	87,026			92,406			88,432	
	61,091	0.708		69,023	0.729		71,748	0.800
	65,064			68,002			57,396	
25.4	22,134		26.5	24,528		35.2	31,085	

Demand	% in Calorie	Demand	% in Calorie	Demand	% in Calorie
25,034	37.0	25,853	33.7	24,802	31.4
252	0.3	237	0.3	136	0.1
264	0.4	302	0.3	234	0.3
17,188	24.4	19,774	24.7	21,278	25.8
(314,743)	62.1	(339,473)	59.0	(341,603)	57.6
345	0.5	470	0.6	215	0.3
8,541	16.7	11,043	19.0	12,121	20.2
33,438	16.1	37,960	16.1	39,868	16.5
2,027	0.7	3,326	1.0	3,312	0.9
4,145	3.9	5,127	4.3	5,514	4.5
(507,902)	100.0	(576,410)	100.0	(593,095)	100.0
80,595	12.7	89,720	12.5	88,897	12.0
8,710	8.2	9,690	8.1	9,796	7.9
939	0.4	1,140	0.4	1,085	0.4
(108,162)	21.3	(120,568)	21.0	(120,309)	20.3
5,836		6,238		6,707	

Total energy consumption of the steel industry

The total energy consumption of the Japanese steel industry is as shown in Table 3. According to the table, the energy consumption per ton of crude steel was lowest, 5.73 million kcal, in 1967. This was caused probably by the reduction in the fuel consumption of blast furnaces and the modernization of various facilities including converters. However, energy consumption has shown a rising tendency for several years. The main reasons are (1) the increase in electric power consumption as a result of the strengthening of dust collection for environmnetal improvement and (2) the increase in energy consumption caused by the raising of the degrees of raw materials and fuel treatments, and of products processing.

Energy consumption of the steel industry by sections

The energy consumption of the steel industry classified by sections is shown in Table 4. Of the various sections, the ironmaking, rolling and tubing, and ferroalloy sections indicate higher rates of increase than the rate of increase in total energy consumption. The increases for these sections were caused mainly by factors (1) and (2), and are responsible for the increase in the total energy consumption.

In Japan, as shown in Table 5, the scale enlargement of blast furnaces has already advanced in the ironmaking field and the production rate of converter steel has reached 80% in the steelmaking field so that large reductions in energy unit consumption cannot be expected in these fields. On the contrary, there is a strong possibility of the antipollution cost to increase in the future. Therefore, nowadays when energy crisis and environmental pollution are the greatest problems, the improvement of thermal efficiency and the prevention of environmental pollution accompanying energy consumption are required (which must of course be carried on along with steel production).

TABLE 4

Energy Consumption of the Steel Industry by Sections

Year	Total	Iron-making	Steel-making	Rolling and Tubing	Power generation, Ferroalloy, and boilers		Forging and casting	Others
				Consumption (10^9 kcal)				
1960	143,588	59,872	20,859	21,655	14,587	10,020	3,497	13,119
1965	264,055	123,445	24,840	45,433	28,273	13,883	3,697	24,484
1970	580,607	287,856	32,760	102,985	53,946	33,003	6,418	63,639
				Component rate (%)				
1960	100	41.7	14.5	15.1	10.2	7.0	2.4	9.1
1965	100	46.8	9.4	17.2	10.7	5.3	1.4	9.3
1970	100	49.6	5.6	17.7	9.3	5.7	1.1	11.0
				Annual average rate of increase (%)				
'60-'65	12.9	15.6	3.6	15.9	14.1	6.7	1.3	13.3
'65-'70	17.0	18.6	5.7	17.8	13.8	18.9	11.7	21.0

TABLE 5

International Comparison of Steelmaking Facilities[a]

Process	Japan	U.S.	Soviet Union	West Germany	U.K.	France	Year of Comparison
Scale enlargement of blast furnaces (ton/day/BF production capacity)	2,723	1,465	1,644	1,039	845	672	1969
LD converters							
Quantity	88	76	37	35	20	16	1971
Production capacity (10^3 tons)	101,000	70,990	25,400	35,400	13,860	8,940	1971
Production rate (%)	80.0	53.2	17.2	61.8	38.7	37.1	1971
Continuous casting equipment							
Quantity	63	58	73	32	14	10	June 1970
Number of strands	143	152	173	94	37	21	"
Ratio to total crude Steel production	17.0	unpublished	unpublished	11.0	2.1	3.2	1972

[a]Source: Iron and Steel Statistics Manual.

TABLE 6

Biannual Consumption of Coke, Heavy Oil, and Electric Power by Japan's Steel Industry[a]

USE	FY 1965		FY 1970		FY 1972	
Crude steel production (1,000 MT)	41,296		92,406		102,972	
Coke (1,000 MT)						
For sintering	1,398	8.6%	4,229	11.2%	4,859	12.1%
For ironmaking						
Blast furnaces	14,116	87.3	32,370	85.3	34,219	85.1
Other furnaces	184	1.1	242	0.6	165	0.4
For others	482	3.0	1,087	2.9	974	2.4
TOTAL	16,180	100%	37,928	100%	40,217	100%
Heavy oil (1,000 MT)						
For blast furnaces	1,045	21.3%	3,408	30.9%	4,448	34.9%
For open hearth furnaces	684	14.0	312	2.8	194	1.5
For rolling	1,855	37.9	4,180	37.9	4,866	38.2
For power generation and boilers	733	15.0	1,637	14.9	1,806	14.2
For others	578	11.8	1,489	13.5	1,421	11.2
TOTAL	4,895	100%	11,023	100%	12,735	100%
Electric power (million kWh)						
For ironmaking						
Blast furnaces	575	2.5%	1,103	2.2%	1,885	3.3%
Other furnaces	742	3.2	585	1.2	392	0.7
For ferroalloy	4,315	18.5	9,932	20.1	9,843	17.2
For steelmaking						
Open hearth furnaces	321	1.4	134	0.3	77	0.1
Converters	525	2.3	1,176	2.4	1,605	2.8
Electric furnaces	5,451	23.4	9,541	19.3	11,465	20.1
For rolling	6,458	27.7	13,907	28.1	16,057	28.1
For others	4,915	21.0	13,108	26.4	15,788	27.7
TOTAL	23,302	100%	49,486	100%	57,112	100%

[a]Source: "Iron and Steel Statistics Monthly Report," published by the Ministry of International Trade and Industry.

Major energy consumption of the steel industry by uses

Table 6 shows the consumption of coke, heavy oil, and electric power classified by use. The energy consumption itself increased with the crude steel production. In 1972, a total of 40.2 million tons of coke was consumed, of which 85% was for blast furnaces and 12% for sintering.

The heavy oil consumption for the year was 12.7 million kiloliters, and the consumption for blast furnaces and that for rolling (heating furnaces and soaking pits) accounted for 35 and 38%, respectively, totaling above 70%. The total consumption of electric power was 57.1 billion kWh, of which 28% was for rolling, 20% for electric furnaces, and 17% for ferroalloy.

THE ENERGY SYSTEM IN AN UP-TO-DATE PLANT

The energy system will be explained by taking the example of Kimitsu Works, a highly modernized steelworks of Nippon Steel Corporation. At this new works, the enlargement of the production scale and the increase in the production speed have been pursued in consideration of the land utilization efficiency and equipment investment efficiency. At the same time, measures to raise the energy efficiency have been sufficiently taken in the operation of this works. The plant layout represented in Fig. 1 was decided from such a point of view.

All on-line (AOL) system

In constructing this works the highest degree of computer utilization possible was sought. As a result of this endeavor a perfect on-line real time system, known as the "Kimitsu AOL System," was developed and applied to the integrated steelworks for the first time. An outline of the system is given in Fig. 2.

Although its details are not described here, this system is believed to be able to adequately satisfy the requirements of customers in all aspects of business including quality, cost, and

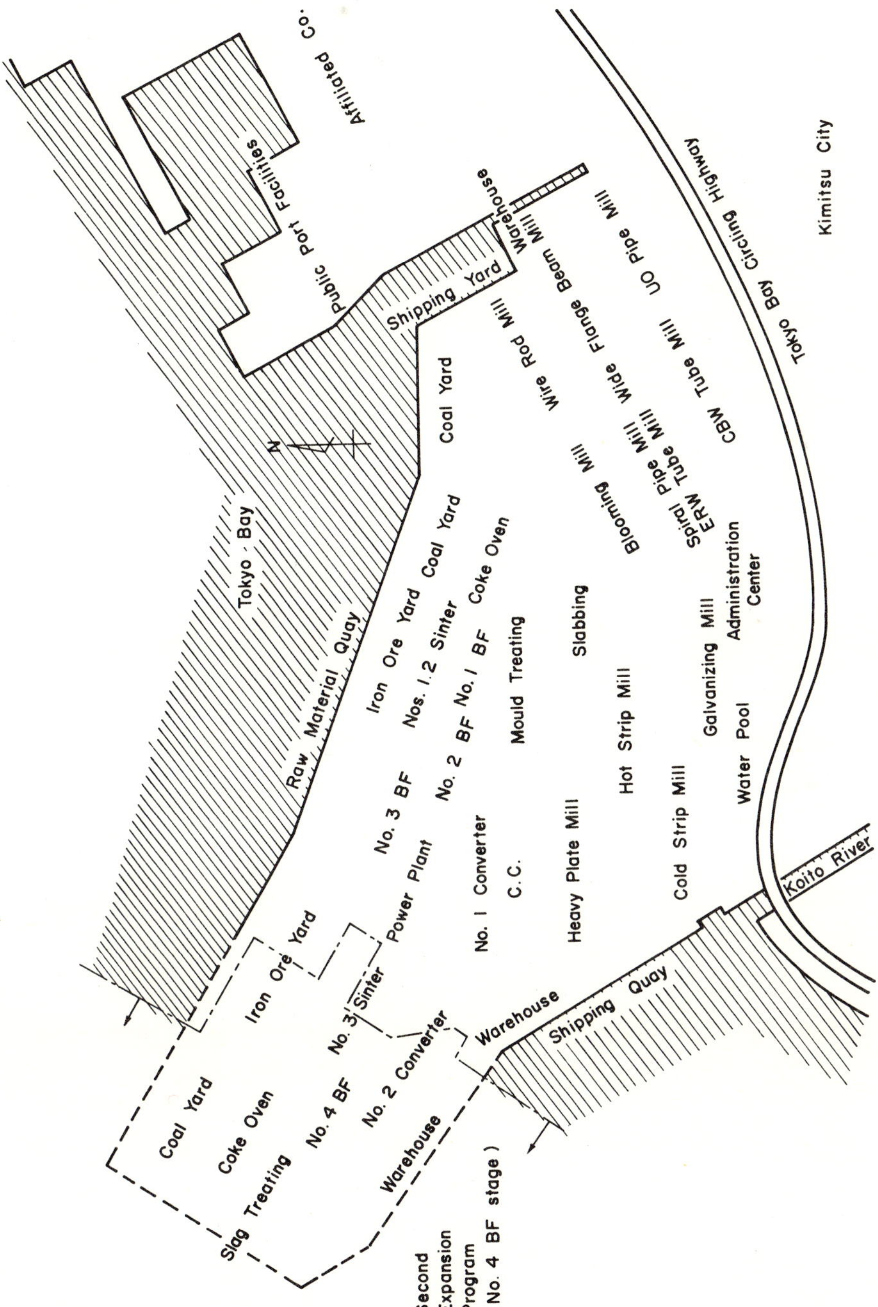

Fig. 1. Plant layout of Kimitsu Works.

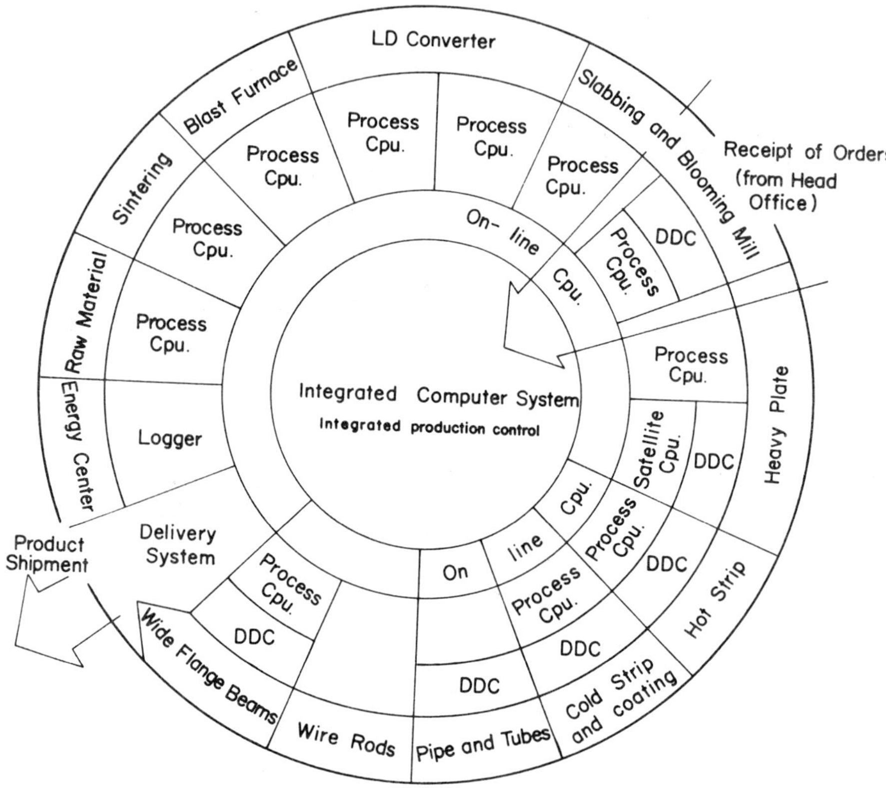

Fig. 2. All on Line (AOL) System at Kimitsu Works, NSC.

time of delivery. This is accomplished by controlling the entire process from receipt of orders to shipment of finished products all in a quick and integrated manner with the computer networks.

Features of the energy system

Adoption of the corporative thermal power system and electrification of the blast furnace blowers

In order to enjoy the advantages of home-generation of electric power and to lighten the absolute values and instantaneous variations of power requirements, which were increased particularly by the scale enlargement of rolling mills, corporative thermal power plants were constructed in the works and large blowers for the blast furnaces were electrified as its base load. This has enabled the installation of a large power plant, and the home-generated power from the corporative thermal power plants has exceeded 90% of the total power consumption.

The oxygen plants for the converters and blast furnaces serve to reduce the oxygen cost by using low-cost electric power while fulfilling its role of base load together with the large blowers.

Effective utilization of gases

Although coke oven gas and blast furnace gas are partly transferred to the heating furnaces and others, the blast furnace gas in particular is used in combination with heavy oil as a major fuel for the corporative power plants to supply low-cost electric power. Furthermore, the waste gas exhausted from the converters is recovered by the OG system developed by Nippon Steel and, after being dedusted, is used together with COG and BFG. When a plant is out of operation, it is possible to consume the surplus gas in the corporative power plants and transfer the saved heavy oil to storage. Therefore, the corporative power plants serve as buffers for the gas balances, which may be disturbed by process variations in the works, and also enable the abundant use of gas fuels.

Efficient use of water

The cascade method is used for the water supply systems for both fresh water and seawater, so that the supplies and returns can be aptly controlled. The rate of recovery of used water is above 93%, and the water supplies are not affected even by abnormal water shortages.

Operation of the energy distribution equipment

Remote control systems and other modern systems are incorporated in the power receiving and distributing equipment, gas calorie control and distribution equipment, and heavy oil distribution equipment. The so-called energy control of these facilities is centralized by the energy center and largely rationalized by the full utilization of computers. The material flow and the energy flow of Kimitsu Works are illustrated in Fig. 3 and Fig. 4, respectively.

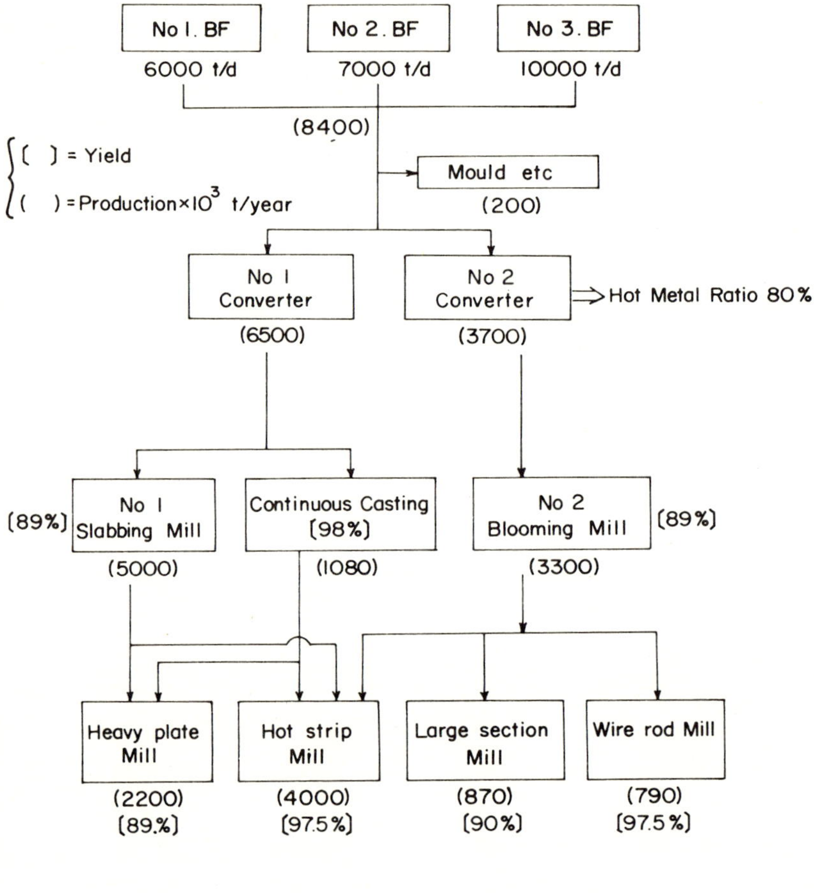

Fig. 3. Material flow at Nippon Steel's Kimitsu Works.

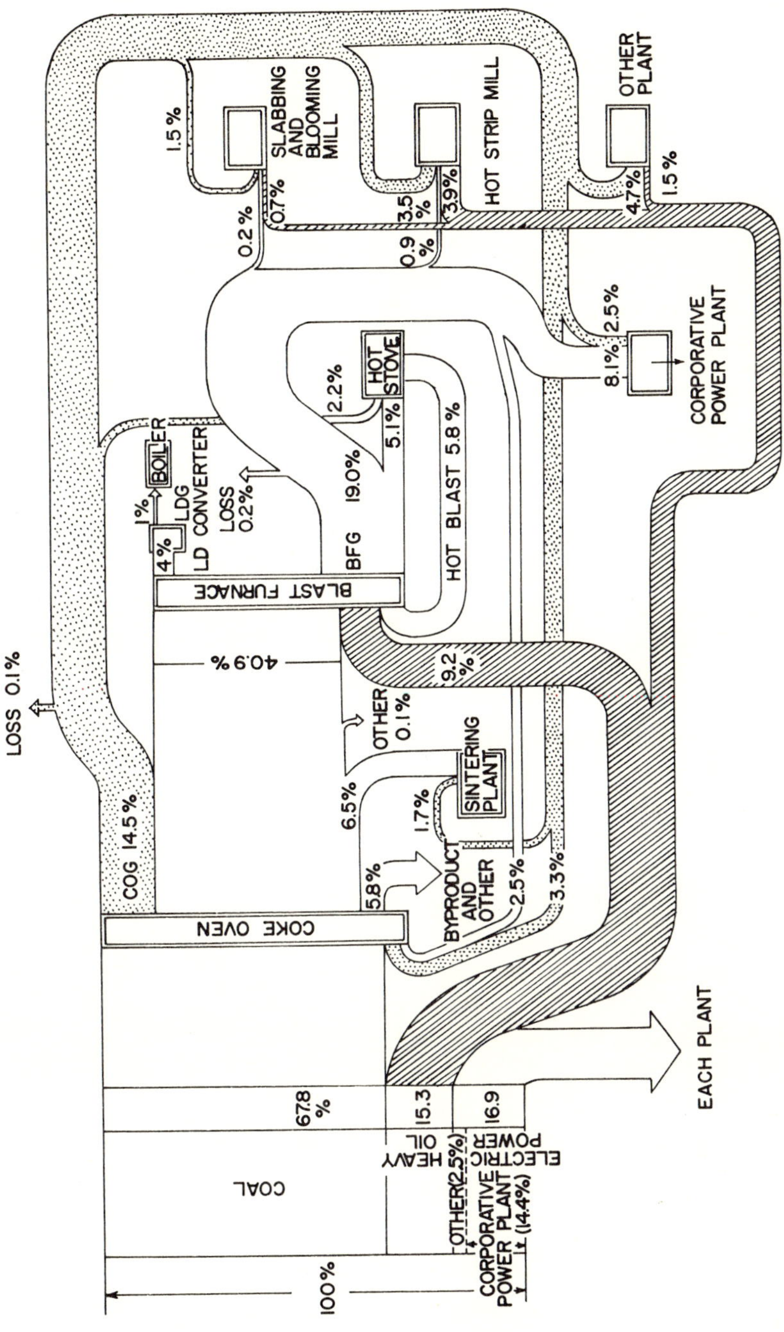

Fig. 4. Typical energy flow diagram of a modern intergrated steelworks (Kimitsu Works, NSC, 1970).

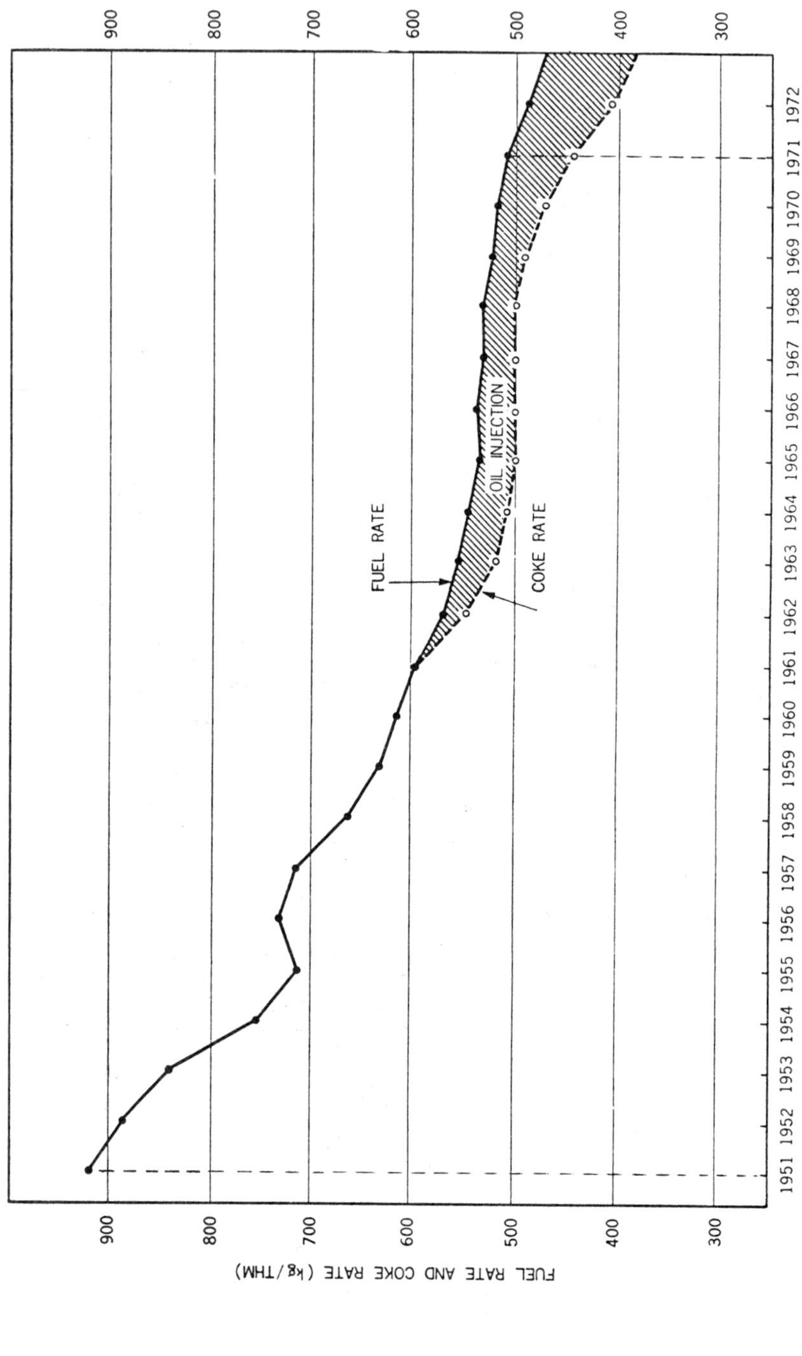

Fig. 5. Trend of fuel rate and coke rate of blast furnaces in Japan.

REDUCTION OF BLAST FURNACE FUEL CONSUMPTION AND TECHNOLOGICAL DEVELOPMENT

The weight of costs at an integrated works is very high, with about 40% of the ironmaking cost occupied by the cost of coal. For this reason, like the improvement of yield, the reduction of the blast furnace fuel rate, especially the coke rate, is one of the most important means for lowering the cost of the steelmaking process.

Figure 5 represents the transition of fuel rate for blast furnaces in Japan. As shown in the diagram, the fuel rate decreased from 915 kg/THM to 508 kg/THM, and the coke rate decreased from 915 kg/THM to 447 kg/THM in the past 20 years. The following may be mentioned as technical factors in this improvement.

1. Improvement of furnace burden properties
2. Use of higher blast temperature
3. Fuel injection through the tuyers
4. Establishment of equipment and operation techniques for stabilized furnace operation
5. Manufacture of high-quality coke

These factors are shown in more detail in Fig. 6.

This figure indicates in time series the progress of development and application of these techniques, and the spread of their wide-ranged use in the past 20 years. It is most important to individually incorporate effective techniques in the highly complex process combining the preparation of ore and coal, sintering, pelletizing, coke oven operation, and blast furnace operation. It is also important to confirm the results of the newly adopted techniques and their economic features to see whether or not they are as originally intended. This will be useful for ensuring the future development of related techniques.

Figure 7 compares the means of lowering the fuel and the degree of influence of these reductions achieved recently by Nippon Steel's blast furnaces, with that of ten years ago.

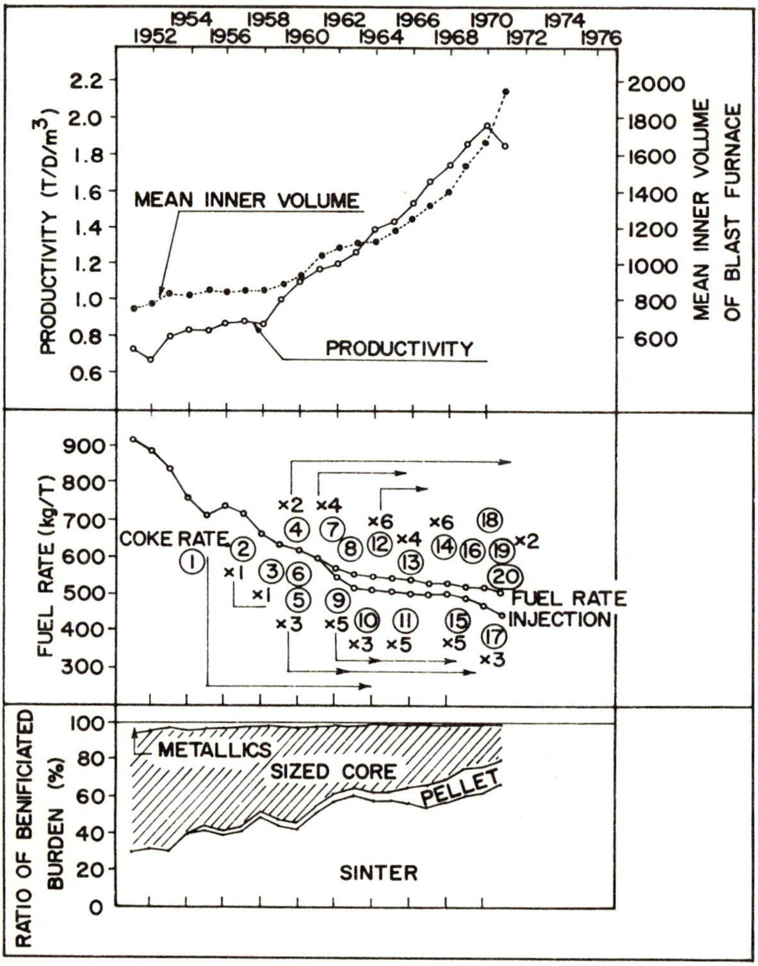

Fig. 6. Trends of technical development and its application to the iron making contribution toward improvement of blast furnace operation in Japan.

Fig. 6. (continued)

1. Strengthening of ore sizing
2. Development of self-fluxing sinter 1
3. Change to self-fluxing sinter 1
4. Construction of large blast furnace 2
5. Application of fuel injection at tuyere 3
6. Application of combined blast
7. Increase of blast temperature 4
8. Wide spread of oxygen converter
9. Application of high top pressure operation 5
10. Increase of fuel injection 3
11. Spread of high top pressure operation 5
12. Use of imported pellet 6
13. Application of high blast temperature hot stove 4
14. Increase of pellet ratio 6
15. Construction of super high top pressure blast furnace 5
16. Spread of oxygen-enriched blast operation
17. Extreme increase of oil injection 3
18. Application of on-line computer control operation
19. Construction of 10,000T/D blast furnace 2
20. Application of coal briquette blend coking process

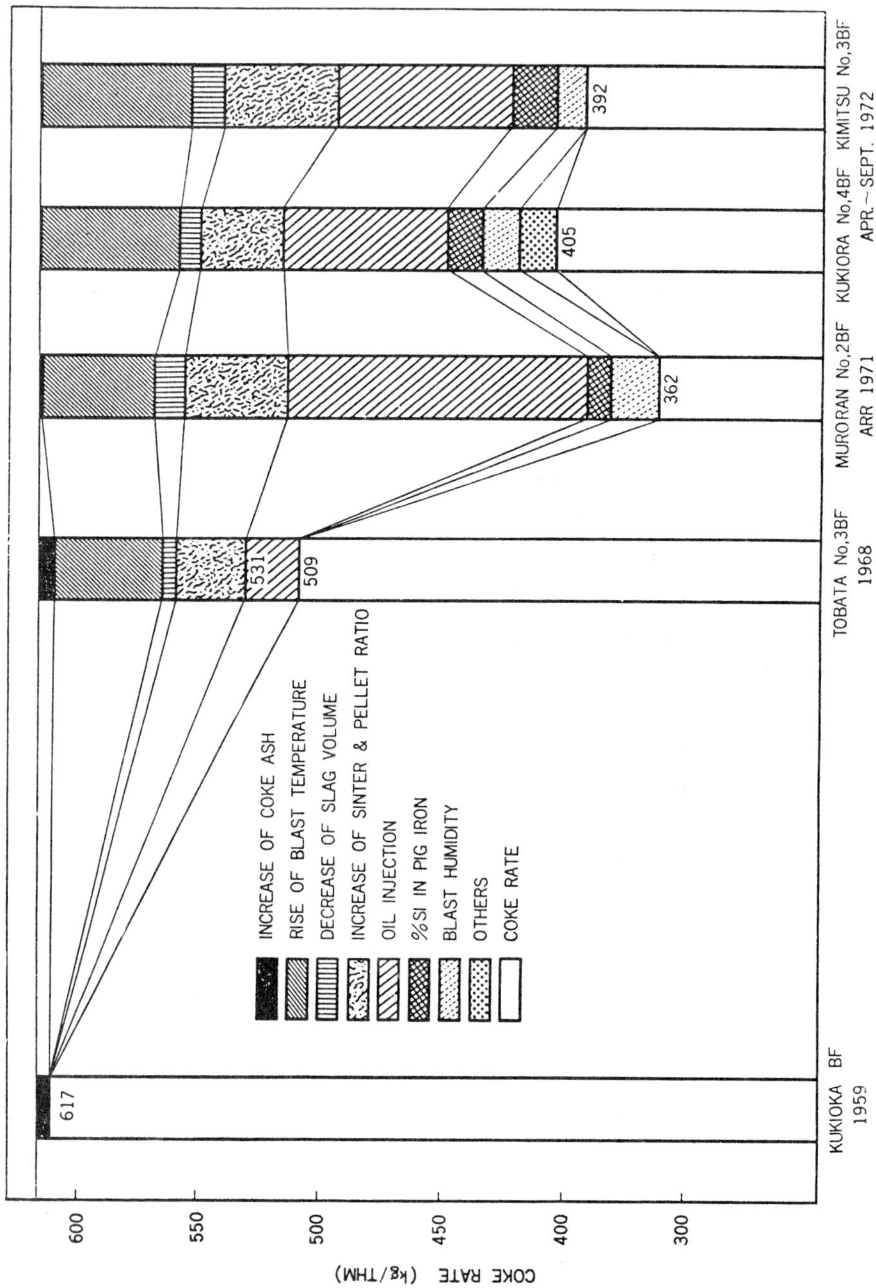

Fig. 7. Contribution of various factors in blast furnace operation to coke rate (kg/THN) in NSC.

TABLE 7

Coke Rate (kg/THM) in Blast Furnace[a]

Country or Region	Actual						Estimated	
	1965	1966	1967	1968	1969	1970	1975	1980
U.S.A.	665	647	641	629	630	630	570	525
Canada	545	547	540	520	514	509	490	480
E.E.C.	701	660	628	611	594	583	500	475
U.K.	680	674	656	656	650	628	540	500
Austria	617	589	549	519	502	507		
Spain	906	938	884	827	796	745	570	550
Sweden	537	525	542	528	540	536		
Japan	498	497	496	499	492	474	400	380
Australia	629	620	613	586	589	601	570	560
Nine Country weighted average	651	628	610	597	586	574	500	469

[a]Source: IISI Official Report, "Report of the working group on coking coal on the future prospects of coking coal 1972."

In these three blast furnaces with high fuel efficiency, high-quality coke and ultrahigh-quality self-fluxing sinter are being used at high rates. Sinter is used 100% in Kimitsu No. 3BF. High temperature blowing, oxygen enrichment, and high top pressure operation are also being used in these blast furnaces. When these functions are explited through the modernization of facilities, it will become both possible and practical for any blast furnace to achieve a fuel rate of 450 kg/THM and a coke rate in the range of 360-400 kg/THM. The coke rate transition in foreign countries during a five year period, and its future prospects, are shown in Table 7.

From the point of view of fuel resources, the worldwide reduction of the coke rate in blast furnaces is of great significance. It is considered that all countries should make further advances toward this end by mutual cooperation.

ANTIPOLLUTION MEASURES IN THE JAPANESE STEEL INDUSTRY

Estimation of the future steel demand and the requirements of clean industry

The estimation of the scale of energy demand in Japan's steel industry is related to the scale of future crude steel production. The Iron and Steel Section of the Japanese government's Industrial Structure Council has published the following estimations.

As for the domestic demand, the average annual growth rate for the whole of the 1970s will prove to be lower than 10.8%, the figure for the latter half of the 1960s, but the demand will reach 110-120 million tons in 1977 (the average annual growth rate for 1972-1977 being 8.5-10.4%). Exports are expected to reach 35 million tons in 1977. Consequently, Japan's crude steel production for 1977 will be 145-155 million tons. (The average annual growth rate for 1972-1977 is estimated at 7.1-8.5%, though that for the latter half of the 1960's was 12.1%.) In order that the Japanese steel industry can continue its growth rate, it must emerge from its present state into a clean industry. First of all, the industry must urgently take adequate steps to solve environmental pollution problems.

Starting for clean industry

The steel industry of the 1970's has progressed toward clean industry, tackling environmental pollution problems by dint pooling its talents and exerting its utmost to attain to legal standards set for environmental quality. And furthermore it has tried to lower the environmental pollution load to levels below those permitted by law. Some antipollution measures to be taken by the steel industry are described below.

1. The environmental quality standard for SO_2 must be reached and maintained by decreasing the volume of exhaust SO_2 at each steelworks by 60% in five years. This will be done by further reductions in sulfur content of raw materials and fuels, desulfurization of COG, and desulfurization of stack gas at sintering plants dependent upon increase of production capacities and characteristics of plant localities.
2. The NO_2 exhaust volume must be decreased by improving the boilers and other equipment of the steelworks. After this is accomplished the legal standard must be reached and maintained by establishing a firm technical development structure through the total efforts of industry, thereby developing fundamental prevention and removal techniques with the guidance and aid of the government.
3. As for waste water, the present water recirculation (rate of recovery of above 75%) must be improved by the adoption of closed systems to eliminate water contamination problems and heighten the effect of economizing water resources.
4. Course particles, dust and soot exhaust volumes, must be largely decreased by strengthening the self-sprinkling systems in raw material yards and employing dry dust collectors for coke ovens, blast furnace site dust collectors, and dust collectors for steel furnace throats and for buildings.

In addition to these efforts, an effort is being made to create new natural environments by using ecological techniques such as surrounding steelworks with green belts, thus going beyond the scope of mere prevention of industrial pollution.

ENERGY MEASURES OF THE JAPANESE STEEL INDUSTRY

Among resource and energy problems, especially in the field of energy problems, there are numerous problems which cannot be solved by the steel industry alone. They must be solved by using integrated countermeasures organized along the lines of the Japanese industrial structure as it should be, and economic operations in the 1970s.

However, it is imperative for the steel industry to take the initiative in fulfilling its responsibility as an industry consuming resources and energy by taking the following measures.

Petroleum group fuels for steelmaking

Assuming that according to estimates, the crude steel production is 130 million tons in 1975 and 150 million tons in 1977, and that the unit consumption of heavy oil to be injected in blast furnaces is 85 liters/THM, and because the present tendency to switch from coke to heavy oil may be expected to continue for the time being, the estimated quantities of heavy oil consumption are as shown in Table 8.

TABLE 8

Estimated Fuel Consumption of the Steel Industry

Industrial process breakdown		1975	1977
Crude steel production: 10^6(A)		130.0	150.0
Blast furnace pig iron production: 10^6 tons (B)		104.0	120.0
Consumption of heavy oil and other fuels			
For blast furnaces 10^6 Kl (C)		8.9	10.2
For rolling and others 10^6 Kl (D)		11.0	12.8
Total 10^6 Kl (E)		19.9	23.0
Unit	C/B: 1/THM	85.0	85.0
Consumption	D/A: 1/THM	85.0	85.0
	E/A: 1/THM	153.0	153.0

The securing of heavy oil, particularly low sulfur content heavy oil, to meet the estimated consumption is anticipated to become increasingly difficult. Although means of reducing the sulfur content of fuels are to be studied on the national level, the steel industry should consider that it is its responsibility to substantially cut the sulfur contents of steelmaking fuels by taking the following measures.

1. As blast furnaces have technical characteristics of mammoth desulfurization equipment, technical studies must be hurried so as to enable the use of high sulfur content heavy oil (S, about 1.6%) in blast furnaces for the purpose of saving low sulfur content heavy oil.
2. Efforts must be exerted to ensure the low-cost and stabilized procurement of substitute fuels like IPG and naptha within the limits of the entire supply-demand balance in Japan.
3. The steel industry must guarantee its long-range purchase of desulfurized heavy oil from the petroleum industry, and must also cooperate with the petroleum industry by clearly defining the necessary sharing. Thus petroleum industries can carry out the direct desulfurization of heavy oil , heavy oil cracking, and desulfurization of heavy oil by gasification all in a systematic and smooth manner.

Electric power

The future electric power demand of Japan's steel industry, as shown in Table 9, will be greatly lowered from its present level in terms of average annual growth rate. Nevertheless, the volume of demand will be increased to 85.9 billion kWh in 1975 and 92.6 billion kWh in 1977.

Since the construction of new power plants is at present extremely difficult because of environmental pollution issues arising at the sites, the power reserve rate will lower until, at worst, it will become minus if plant construction is not realized, causing deficits of supply against the total demand in 1977. In order to counter this situation the following measures must be taken by the steel industry.

TABLE 9

Estimated Consumption of Purchased Power and Home Generated Power of the Steel Industry by Manufacturing Modes 10^9 kWh

Manufacturing Mode	FY 1972	FY 1975	FY 1977	Annual average rate of increase	
				1972-1975	1972-1977
Integrated steel works					
Purchased power	12.52	16.76	18.40		
Home generated power and others	20.47	28.77	31.60		
Sub total	32.99	45.53	50.00	11.2	8.6
Open-hearth and electric furnace plants and others					
Purchased power	29.50	39.00	41.27		
Home generated power and others	1.20	1.39	1.39		
Sub total	30.70	40.39	42.66	9.7	6.9
Total consumption of power of the steel industry					
Purchased power	42.02	55.76	59.67		
Home generated power and others	21.67	30.16	32.99		
TOTAL	63.69	85.92	92.66	10.5	7.8
Crude steel production(1,000 MT)	10,972	13,500	15,000	--	--

Note: The column for "Home generated power and others" includes corporative thermal power.
Source: Data published by the Ministry of International Trade and Industry.

Promotion of home generation or corporate generation thermal power

Integrated steelworks have been making efforts to minimize the purchase of power from electric power companies by effectively utilizing part of the BFG produced in the works as fuel for home generation plants or corporate power plants. This effort must be increased and continued in the future with adequate attention given to antipollution measures.

Increase in the cost of electric power and its effects on plant operation

Substantial increases in the cost of electric power are considered unavoidable owing to the decline of the financial environment which surrounds the business. Such factors include the recent sudden rise of fuel costs, the increase of circulating fixed charges with its root in the necessity of installing antipollution equipment, and land procurement problems. Therefore, it will become highly important for steelworks, especially electric furnace plants which depend greatly on purchased power, to use power effectively by consuming or saving it in a way suited for the load patterns of the power plants.

Coking coal

The consumption of coking coal in the steel industry is estimated at 69.7 million tons for 1975 and 80.4 million tons for 1977, as shown in Table 10.

TABLE 10

Estimated Consumption of Coking Coal in the Steel Industry 10^3 MT

Use or source		
Crude steel production	130,000	150,000
Blast furnace pig iron production	104,000	120,000
Domestic coal	8,000	8,000
Imported coal	61,700	72,400
Total consumption of coking coal	69,000	80,400

Note: Coke rate: 400 kg/THM Coal rate: 670 kg/THM

In the past, the long-period stabilized securing or reservation of iron ore and coking coal, the main raw materials for steel production, was the greatest problem to be solved by the rapidly-growing Japanese steel industry. The stabilized securing of coking coal was enabled by the following:

1. Efforts to obtain Australian and Canadian as well as American coal
2. Conclusion of long-period contracts with foreign suppliers
3. Securing of transportation power and reduction of freights by adoption of large coal-carriers

Under the existing circumstances, no problems are expected with regard to coal supply for the next two or three years. However, the international environment of iron resources is presently undergoing a marked change. The mere continuation of the above measures with long-period purchase contracts as their basis is no longer sufficient and is already not without new problems. As measures to counter such a situation, the following must be mentioned.

1. The further dispersion and expansion of sources must be considered in the future to facilitate the securing of required quantities and the stabilization of supplies.
2. The coking coal mining conditions will become more and more unfavorable in the future and the increase of digging costs will bring about steep rises in coal price. Therefore, utilizing extant low-grade coal is a measure worthy of note also from the viewpoint of the effective utilization of natural resources. A means of realizing this end, such new techniques as the formed coal process and the formed coke process are presently under study. Much is expected from the results.

OUTLOOK OF IRONMAKING METHODS

Transition and prospects of the blast furnace process

As stated already, various technical improvements have been made in the present blast furnace process. Figure 8 shows transitions which are expected to take place in the blast furnace process from the viewpoint of fuel reduction and of environmental pollution.

For blast furnaces, ultrahigh top pressure operation, heavy oil volume injection, and reduction gas injection may be considered as future techniques. For cokemaking, mixing of coke with formed coal, charging of dried coal and charging of preheated coal, dry quenching and continuous cokemaking, like the formed coke process in the place of room furnace cokemaking, will be employed. As for iron ore, the use of oxidized pellet and reduced pellet will be increased. Also, as a future ironmaking process, direct ironmaking by new type low shaft furnaces and electric furnaces will be adopted by direct use of reduced briquette or pellet and iron ore.

Introduction of nuclear power into ironmaking

In the development of the blast furnace process and the actual adoption of the direct ironmaking process, nuclear energy will certainly come to have great importance in Japanese life and industry. Nuclear energy will be introduced also in steelmaking. Figure 9 represents the patterns of its introduction into ironmaking.

The direct-reduction electric furnace steelmaking process, regarded as the standard process for nuclear energy steelmaking, is still unable to compete with the blast furnace - converter process in productivity and cost under the present conditions and also involves numerous items requiring further investigation and development. In Japan, technical and economic studies have been conducted by the Nuclear Energy Section of the Japanese Iron and Steel Association, on the introduction of nuclear energy into the steel industry for the past five years.

POSTSCRIPT

It is believed that the superiority of the blast furnace-converter process in the steel industry will continue into the future. We consider that, while we should pursue the possibility of converting the present process into new processes, we should also exert greater efforts at the present time toward the effective utilization of coal

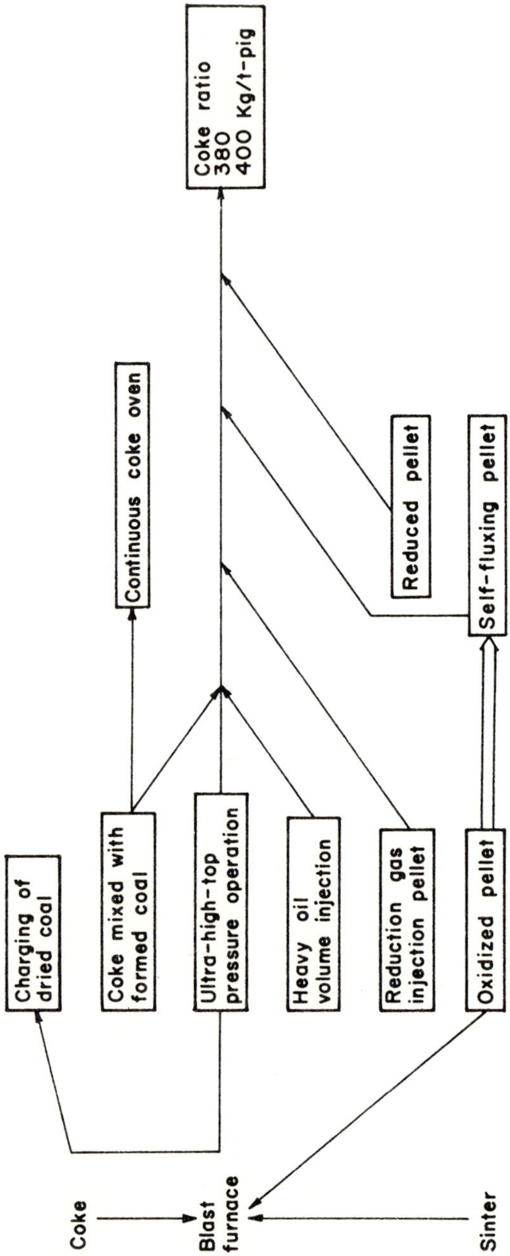

Fig. 8. Transition in the blast furnace process and its future in steelmaking.

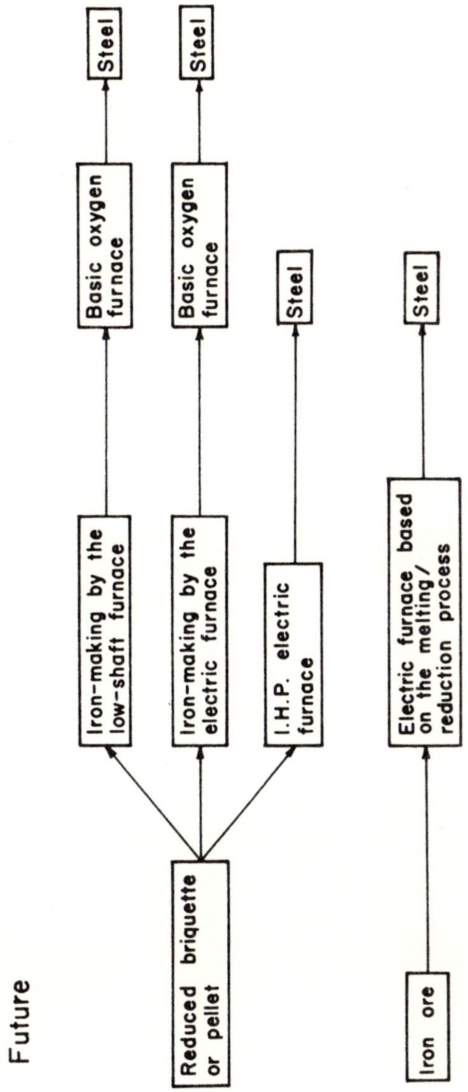

Fig. 8. (continued)

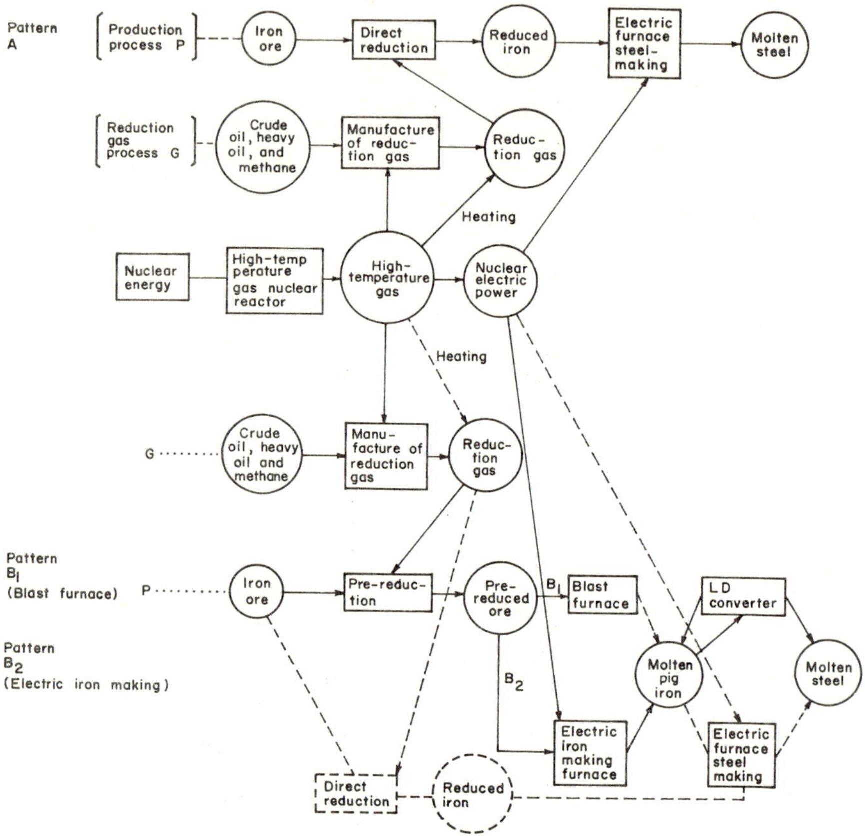

Fig. 9. Introduction of nuclear energy into ironmaking schematic representation of the process patterns. (A) Direct ironmaking via electric furnace steelmaking; (B) via molten pig iron (B_1) blast furnace process (B_2) electric ironmaking furnace process.

and major resources for steelmaking. Although the prevention of environmental pollution must, of course, be considered in this connection, these efforts should prove to contribute to the improvement of the energy economy from the standpoint of the whole of Japanese industry.

INDEX

Numbers in parentheses are reference numbers and indicate that an author's work is referred to although his name is not cited in the text. Underlined numbers give the page on which the complete reference is listed.